AF691822

COMMONPLACE MYSTERIES

COMMONPLACE MYSTERIES

ELROY BODE

Texas Western Press
THE UNIVERSITY OF TEXAS AT EL PASO

Copyright © 1991
Texas Western Press
The University of Texas at El Paso
El Paso, Texas 79968-0633

All rights reserved. No part of this book may be used or reproduced in any manner without written permission from Texas Western Press, except in the case of brief quotations employed in reviews and similar critical works.

First Edition
Library of Congress Catalog Card No. 91-065176
ISBN 87404-219-4

The paper used in this book meets the minimum requirements of American National Standard for Information Sciences.
∞ Permanence of Paper for Printed Library Materials, ANSI Z.39.48-1984.

Book design by Rebecca Quiñones

Some of the sketches and articles appeared originally in the *El Paso Herald-Post, Nova, Rio Grande Review, Southern Living,* and *The Texas Observer* and are reprinted with permission.

CONTENTS

EL PASO WANDERINGS

ON SUNSET ROAD

THE HILL COUNTRY

EL PASO WANDERINGS

Small Swept Yards and Green Porch Posts

Day is ending on Magoffin Street.

I walk past red-brick houses with red-brick chimneys — past small homes, small yards, small boys, small corner grocery stores. Women holding babies talk to women with other babies as the light fades and night coolness comes on.

They are still here on Magoffin, Myrtle, San Antonio, Olive, St. Vrain: houses left over from the turn of the century, houses that were once the heart of town. They sit side by side with their attic windows, peaked roofs, and green porch posts. Some of the yards are grassless — just sandy dirt — but they are swept clean as a plate. Others are neat, shaded sanctuaries filled with marigolds and cannas, oleanders and arbor vitae. Pink-flowered 'miguelito' vines — Queen's wreath — are draped like modest shawls over doorways and along fences. Children are everywhere: in doorways, in alleys, on steps, in apartment hallways. They throw a ball, ride bicycles in slow, constant, almost meditative circles. Sisters carry little brothers in diapers, lead two-year-olds along the sidewalk. Boys who are not playing touch football in the street are playing pitcher-and-batter in a cluttered yard: the pitcher throws flattened beer cans and the batter swings at them with a stick. Children call out constantly, and other children answer from a distance — their voices rising and drifting in the late afternoon air like bird wings changed into sound.

After the sun goes down there is a long, suspended moment: intimate, familiar, personal yet strangely distanced, like a scene in a play. A cat crosses a yard, eases over a curb, seats itself beside a parked car. Immobile, tail curled, it awaits no one and eternity. The brick chimneys seem suddenly dramatic and exotic, looming up against the pale western sky. Houses, yards, alleys withdraw into themselves — become private. Men sit in the yellow-lit doorways of auto repair shops, looking out, beer cans in hand. A thin, gray-haired man in an undershirt sits smoking within the dimness of his porch.

After dark the neighborhood network of play, of voices, of lives remains intact. Moonlight, porch light, street lights appear, and the streets shine like the smooth, stretched skin of elephants. A train whines somewhere past Cotton Street, then dies — and the tree frogs resume their rhythmic pulsing. Girls talk on porch steps but keep careful watch as lone cars cruise slowly around corners, their tail-lights like twin messages from Other Places. Here and there a house sits in isolation — sunk beneath the weight of the century, half-buried behind dying palms and chinaberry trees. A telltale bulb glows within: a grandmother still lives in a back room under a rotted roof.

As I stand on Magoffin Street — the scooters, tricycles, plastic Big Wheels still rolling down the dark sidewalks, the invisible children yelling, giggling, squealing — I have a curious vision of the Apocalypse: Someday, yes, the Ultimate Bomb will be exploded by the Ultimate Terrorists. The earth will be rent and cities turned to rubble. The screams will finally end and silence will reign. Then, somewhere, the silence will be broken because a child, in a cave, will see a rock, a stick, a piece of string and out of them he will fashion himself a toy. Playing with it he will become amused, and laughter — a child's laughter — will once more be heard in the land. Other children will hear it and begin to laugh too, and once more, as the children come out of their caves and gather together, neighborhoods will be possible.

Downtown

The fainthearted generally avoid the downtown library area — Missouri, Oregon, Franklin Streets — and to them I say: I understand. It's not Yuppie City. But I still like it. For almost thirty years the library has been my study, refuge, after-hours home — and the surrounding streets my walk-about territory.

Let's say it's 5:30 in the afternoon and I am returning books to the library as well as taking in a stack of school papers to grade. I give a once-over to the New Books shelf on the main floor; then I go downstairs to the basement and find a vacant table among retirees, drifters, out-of-work laborers, afternoon regulars. Most of them are reading; some are looking at pictures — their unmoving hands holding unmoving books and magazines. Men in baseball caps and work clothes read small, Spanish paperback novels. One toothless man — just released, I hear him say, from jail at 4:00 for vagrancy — is now reading Louis L'Armour at 5:45.

A red-haired man in a flannel shirt jiggles his forefinger across lines of an astronomy text as if he were herding doodlebugs toward the margin or maybe witching for water between the lines. (Is he actually reading? I can't tell. He stares into the dark heavens of the text and jabs his broken fingernail into galaxy after galaxy, then — slamming his hands to his temples — he speaks silently, violently into the page.)

Another transient in rubber-tire shoes — just in from Anywhere and Nowhere — has stowed his enormous pack under a table and is pretending to read while he nods away beneath his stocking cap and grizzly beard.

When I come outside the day is cooling and some of the transients are sitting on the low retaining wall, looking vaguely out to the street. Others are sprawled on the grass, a pants leg hiked up to show a scaling, crusted leg. They lie there — asleep, drunk, worn out as the pigeons walk sedately nearby. Kids from the Hotel Linden across the street kick a wadded handkerchief around in a homemade soccer game as college students and retired military men and downtown knockabouts continue to pass in and out of the shadowed library door. Idle taxi drivers, their windows rolled down, read newspapers in cabs across the way. Hookers make their slow rounds along the block.

One of the downtown regulars — who seems to live his entire, marginal life on his feet with a bundle on his back — passes by, his railroad engineer's cap pulled low over his eyes. A small, bearded Rip Van Winkle, he prowls the sidewalks in cracked, curled-up running shoes. He is a connoisseur of sidewalk treasures and keeps a sharp lookout for cigarette butts, broken Bics, crumpled paper cups.

I walk to the plaza and take a seat under the mulberry trees. An intense young evangelist in blue jeans is striding back and forth on the plaza platform, shouting his message to us in Spanish. He preaches, exhorts with his raised Bible, and never stops chewing gum.

A young man in an undershirt slouches into a nearby bench. Sleek-haired, he wears dark glasses, gray work pants, white socks. And tattooes. He has tattooes on his arms, neck, hands — more tattooes than brown skin.

He listens to the evangelist a while, slings himself back to his feet, walks on. Almost immediately his place is taken by a younger man, perhaps 19 or 20, who seems to have emerged from underground or maybe a drainage tunnel. It is as though he has dirtied himself dramatically for a part in a movie: Drainage Man. Befuddled, shaping soundless words, apparently possessing nothing more than the rags on his body, he moves aimlessly from bench to bench and then out of sight.

All the while — as the ice cream vendor jingles his way through the shadows, as women get up from benches and run for their bus — a young Chicano with lovingly shaped black hair sits balanced on the rim of a brick planter, oblivious of his surroundings. One hand rests on an upraised, creased pants leg. He stares straight ahead, a statue in clean white T-shirt, clean black pants, and pointed, shining black shoes. His statue-girl, also in white T-shirt and black pants, stands immobile beside him, gazing past his earlobe into some vague middle distance. Neither moves; neither speaks. He is providing her with the pleasure of his profile and his carefully stylized hair; she is providing him with her dutifully adjacent self. At first they seem afflicted by a strange, late-afternoon paralysis, but finally the girl does move. Her face comes forward like a boat being steered into dock; the young man allows her to place her lips into contact with his; she then moves back, returning to her normal, expressionless pose. It could be that they are in love, but it is hard to tell.

Fall Afternoons in the Upper Valley

The pale sun is out, shadows are lengthening everywhere, there is a moodiness along the streets. I walk in the sun-lit privateness of late November, and it is as though I am moving in the presence of a kind of visual music — as if the papery leaves in the yards are random, elegiac notes drifted down from the mulberry trees.

I walk among dying things, yet death — in November, in the Upper Valley — wears a benign and innocent face. The leaves and weeds and grasses blend into a patina of grays, mild yellows, browns as the landscape — like a terminal patient — burns silently with the cold fevers of autumn.

The light: it hypnotizes. It does not have the cleansing, cauterizing quality of a summer sun, but in touching ordinary surfaces it makes them seem haunting, enigmatic. The plain white side of a house at four o'clock becomes a glowing and indecipherable message — shining with a strange, almost phosphorescent inner radiance.

And shadows: they are suddenly portentous, filled with hidden significance. Shadows of trees stretch across the ground with a primeval strength. The shadows of rocks, of leaves, of weeds are like Dionysian altars — sacred places inviting worship or celebration.

I stop and sit and listen, and it is like eavesdropping on childhood. The sounds of the neighborhood — a car turning on a gravel street, a dog barking, a boy yelling across a yard — are so leisured, so casually spaced it is like the easy stretching of an animal: as if the whole Upper Valley is yawning with Sunday, small-town contentment. . . . Then, for a long moment, nothing moves. The afternoon has merged with the universe. The mild warmth of the sun, the mountains rimming the horizon, the white bolls clustered in the cotton field, the bird sitting in a bush: All is as it should be. The earth is like an eternal Buddha: stilled, cross-legged, smiling.

Going to Juarez

I led an idiosyncratic life. I pursued singular paths.

Or to put it another way: I liked going to Juarez and walking around on a summer day. And the going was as enjoyable as the getting there.

I stopped first for coffee at Victor's Cafe in downtown El Paso. I liked Victor's — maybe it was just the simple, understated invitation of C A F E printed in the window. I took a seat at a side table with its red-checkered cloth. The cook whistled bouncily in the kitchen; the waitress stood in the middle of the room, looking out the doorway at the people passing by. She talked in Spanish to a man seated at a nearby table, having his midday glass of beer. The woman was not a real waitress — that is, she didn't wear a uniform and white shoes and she didn't have a ticket pad at her hip. She was just someone's mother who finally brought me a cup and spoon as she would have brought them to a guest in her own home.

I sipped the coffee, gazed out to the street past the E F A C on the window glass, listened to the whistle-gymnastics of the cook while the small portable fan turned back and forth above the cash register. I sat, blending into the mood of the cafe as easily as the salt shaker, the toothpicks, the coiling cigarette smoke.

After paying for the coffee I eased on past the Greyhound bus station toward the bridge — past the loan shops and dry goods stores; past the Hollywood Cafe and, above it, the bulging eyes of the apartment bay windows that for a century had stared down on the swarming life of South El Paso Street and never blinked.

Before crossing Paisano Drive I went into the Plaza Grocery for a package of gum. The Plaza: a store to dally in for sure, one that still had the tone — the flavor — of the small neighborhood grocery stores of the past. No Muzak there: no swathed, computerized goods lying in state. It was cramped, crowded, narrow-aisled, and it smelled of laundry soaps and brooms and meats from the butcher counter. I could have spent the whole afternoon there, sitting on an apple box and smelling the cleaning compounds.

At the old Colon Theatre I was in the full flow of pedestrians headed toward the bridge. In every block the people, the Mexican people, were walking south – carrying, clutching, clasping, moving with their sacks and packages towards home.

I paid my dime at the toll booth; then, with the others – with old women in scarves and canvas sneakers who had already walked miles and would walk miles more before finally putting their shopping bags down – I started up the incline of the bridge. At the top I paused a moment, letting the others go by. There it was again: the whole hypnotic sweep of river and mountain range and blazing summer sky; the meeting place, the Pass.

In Juarez, on a side street, I watched a woman close a door. I was struck by the care, the consideration, the sense of propriety with which she had opened and closed her old screen door at one o'clock in the afternoon. It was as if the proper functioning of society – indeed, the orderliness of the universe – depended on how well that door was opened and shut.

On Calle M. Ocampo Norte two men walked along pushing a cart of melons. The cart had a microphone attachment and one of the men called out to the neighborhood: *"Sandias . . . sandias!"* They pushed toward a brown and white dog sleeping in the sidewalk beneath a chinaberry tree. *"Sandias!"* the man called, but the dog did not move.

It was an ordinary, hot afternoon. Naked children sat on the cracked, broken sidewalk playing with wooden spools. Women walked in 100-degree heat, holding up their purses to shade their eyes from the sun. Pigeons on the nearby roofs preened and postured to one another. A man strode smartly along in cracked, run-over shoes without socks, holding out his cellophane sacks of limes for sale.

Now and then vans came by, the *maquiladora* workers seated inside as subdued as prisoners. They were two rows of profiles in the windows, staring ahead.

A girl passed – a young mother carrying her baby. Flap, flap, her rubber thongs slapped the sidewalk past the small yards, the doorways, the bougainvillaea here and there, the unpainted fences and gates.

An ice cream man in a straw hat pushed his cart down the sidewalk, up and over the curved bridge of the canal, and on beneath the line of willows where men had gathered to talk. In front of them, on the canal fence, a woman's washing dried in the sun.

Other men, in undershirts, sat in chairs in their doorways, leaning forward on their knees to watch the afternoon. Above them, voices of women drifted out from shadowed, second-story windows. The brown, debris-filled canal water flowed past fading advertisements for Sky sodas and beneath the bar windows where prostitutes had already gathered for the afternoon trade.

On Avenida Juarez I leaned against a store front. Tarahumara Indian women were seated on the sidewalk – palms up, begging. Their sleeping babies were sprawled nearby.

Three Mexican girls walked past and I turned to look at them. They were teenage children with freshly painted and shining lips, high heels, thin bodies, tight jeans. Newly pubescent, they wore men's-style white undershirts with no bras. They stopped for a moment in the middle of the sidewalk – to talk, to laugh, to listen to the drum-and-saxophone combo blaring from an upper-story window across the way.

As they stood there and as I watched them, my first thought was a vaguely patronizing one, something along the lines of: How little they know of *anything*, such girls – how incongruous they seem in a country of such poverty and need.

But what finally struck me was: Just because they live in Juarez, Mexico, why should they wear serious looks, be grave and weighted down? *They are doing the first living they have ever done.* . . . Sure, they were as innocent of the world and its awesome dimensions as reeds along a river bank. They knew nothing of *The Iliad* and ICBM's and the curve of time – but that was their dignity and their strength. They had a right to be young and to be frivolous.

It was as if each girl standing there in her high heels and skin-tight jeans carried a message – wore a placard: Life, wherever it is found, is to be respected.

I could have been leaning against a building in Havana – or Rio de Janeiro or Hong Kong – and the millions, walking there, would be carrying the same message.

Like lilies bursting through concrete, the three painted girls in undershirts were asserting their right to youth, to life — against all odds, in the heat, through the jarring sounds of ragged saxophone music and beside the upturned palms of the poor.

The Man at the Dumpster

As I took a short cut behind the Central Cafe, I passed a young man at a dumpster. He had shaggy black hair and wore a torn black raincoat. He was eating out of the dumpster.

I walked on to the triangular little park next to the Plaza Motor Hotel and turned around — looking back at the man, trying once again to deal with The Question: What does one do when a child on a Juarez street holds out her hand, begging; or a down-and-outer is eating garbage on a hot summer afternoon; or a woman is trudging along at an intersection between long lines of cars with a baby slung across her shoulder and a package of *chicles* for sale in her hand?

Part of me said, logically and properly: "Charity is not the answer. Nothing that I — or anyone — can do today will have the slightest impact on the life of the person in need. It's the entire social structure that needs changing . . . a dime here or a dollar there — or $100 or $1,000 — will not bring about any resolution of the basic problem."

That was reason having its dispassionate say. But another part of me — nudged by conscience and guilt — kept offering its rebuttal: "Hold it! Right now you are in the presence of someone in need. Sure, you'll probably never see him again, and sure, whatever you do will not reshape the framework of society. But are you going to ignore him — pass him by — simply because you cannot change his entire life? Is that a logic to be proud of or is it the shoddiest excuse for a bloodless indifference: 'I've got mine, Jack; let others get theirs the best way they can'?"

I went back toward him. Under his raincoat he wore various shirts, tuxedo pants with a faded silver stripe, black shoes run over at the heels.

"Excuse me," I began, not knowing exactly what I was going to say, "Would you . . . like some money for a meal?"

He kept bringing the garbage up to his mouth and eating from it. Finally he turned his head a little toward me, still chewing, holding a slice of lemon in his hand. I could see the yellow garbage slime across his teeth.

"I have $150," he said. "Yes . . . all I need is my identification papers. They're holding my money at the Acapulco Cafe over there" – and pointed with his lemon slice. He spoke directly to me, reasonably, as if we were discussing quarterly dividends at the El Paso Country Club.

". . . $150. They got it for me. I just need to get my papers."

He leaned forward into the dumpster and flies swarmed above his hands.

I didn't know what to do. He seemed to be a guy with some kind of curious code or dignity: he was obviously rejecting my offer of assistance. I had no idea if his $150 was real or imaginary.

Should I smile and say, "Well, have a nice day"? Should I slip a five-dollar bill into the pocket of his raincoat? I couldn't hand it to him because he was too busy eating.

I walked on toward the plaza, leaving him hunched over the side of the dumpster like a man chowing down at an outdoor lunch counter. He was still there when I turned the corner.

In the Sand Hills

Deborah was seven. She and I had parked by the sandhills on the west side of town and were standing beside the car when a man in a station wagon stopped down the road from us. He opened the door and let out a huge black French poodle who immediately began tearing in our direction, mouth gaped, knobby tail held aloft like some kind of African ceremonial spear. He was about two yards away when the man called out over the top of his station wagon in a strained, ruined voice: "Don't . . . mind . . . Axel . . . he . . . won't . . . bite" – and immediately the dog burst into the space between Deborah and me, disintegrating his charge harmlessly and sniffing at our ankles. I had not moved – had continued to stare at the onrushing dog as though he were some red-tongued demon charging at us across the sands of fate.

The dog sniffed us a while, turned, bounded back to the man in the station wagon. They drove away.

I pulled up the bottom strand of barbed wire and Deborah crawled underneath the fence into the wilderness of sand.

"Oh, Daddy!" she said.

The air was desert-still. Behind us, across the wide El Paso valley, the Franklin Mountains lay like a prehistoric fishback against the sky. In front the small sand hills rose and dipped their way westward.

We walked together in the quiet, losing ourselves from sight of the road. Deborah ran in an excited aimless dance, picking up small rocks, turning around and around, then letting the rocks fly. She found smooth holes in the sand made by desert ants and squatted down to look at them closely, her pigtails swinging forward across her face, almost touching the holes.

She ran up little hills and found pieces of twisted dry wood and stood there with them in her hand while she looked out at the scattered green bushes and the twilight and the sand. Then she raised her shoulders, sighed a little, and said, "Daddy, this is *nice*."

Poor Hammond

I didn't know Hammond well. I had met him at a friend's party, knew he could sing and was interested in acting, came across him from time to time at movies and on downtown streets. He was a handsome fellow and he knew it: I always seemed to find myself talking to his profile.

I saw him most frequently through a window of the public library. I would be inside, reading, and he would go striding past in the late afternoon. It seemed to me – watching Hammond during those brief window intervals – that the phrase had been thought up especially for him: "a fine figure of a man." He was certainly that. Tall, lean, broad-shouldered, looking well groomed in his dark suit, he swung his briefcase with the casual grace of an Ole Miss halfback who had finally entered insurance underwriting. He seemed like a typical young executive headed home toward evenings of young executive fun. I didn't know, of course, that he was eager to kick over the traces – that already his tie was beginning to stay loosened at the neck.

I saw him briefly the weekend before he left El Paso for show business. He had exchanged his suit for a faded red shirt and blue jeans – and a black skull cap to keep his dark hair sleeked down. He was headed east in his Chevy truck; he was going to end up on Broadway. He had $40 in his pocket and a rope and saw in his truck. He said that every town has a house or two where trees need to be trimmed. He had done it before: trimmed and treated trees for a couple of hours and picked up enough gas money for another hundred miles or so.

So he left: Hammond, the tree-trimmer and the ex-executive who had worn his suits custom-made to fit his matinee-idol slimness; who had cultivated a thin, Douglas Fairbanks moustache to set off his flashing Douglas Fairbanks teeth. He drove away in his truck to follow his star.

Poor Hammond. He never got farther than Pine Bluff, Arkansas.

It was a year or so after he had gone that I brought up his name to our mutual friend. "Hammond?" he said, "I thought you knew. . . . He fell out of a tree at a woman's house in Arkansas and broke his neck. She was in the bedroom and Hammond was singing her some opera song or other. He was probably drunk but they said the limb broke. Poor guy, he really did have a good voice."

Alien

One morning in mid-August I decided to take a stroll along Rim Road. It was impressive, as usual – the sweep of streets-and-sky that blended so easily into both time and space. I stood for a long while at the rim's edge, trying to take it all in: two cities sprawled in the sun, spread from horizon to horizon, century to century. . . . My, my, I thought – looking out, looking down – such a place of perspectives, this El Paso; such a city of hillsides and vistas, of flatlands and heights, of neighborhoods that climb and wind and slope. . . . Yessir, a fine place, always interesting to the eye.

I turned away and began walking along the streets north of Rim Road – Blacker, Hague, Campbell. I passed homes of the well-to-do with their smooth hedges and manicured lawns. A quiet neighborhood on a quiet mesa, it was like a museum of tastefully maintained houses and orderly lives – deserted, now, at ten o'clock except for occasional yard men and maids. Down the street a man in a straw hat and blue bandanna hanging down his neck was cutting the grass. On a back porch a young woman in a white apron slowly dusted a rug. Locust sounds soared and then fell dramatically in the trees. A mockingbird sailed down on graceful, outstretched wings and began to sport about in the shining grass near a water sprinkler.

At an intersection I stopped and stared – I couldn't help it. I wanted to absorb the sights of this summer morning into the very tissues of my body. I looked west, and the homes dipped in a smooth, rhythmic line as the Juarez mountains slowly rose in the distance – making a neat and pleasing balance. Overhead, the sky was a cloudless blue saucer, swept clean by an early-fall breeze that kept bathing my face and making a tall sycamore on the corner move its leaves in a continuous sycamore dance. Tree shadows had spread themselves across deep-green yards and into the streets.

Standing there, I couldn't resist the temptation to put something down on paper – to record the moment and the place. I pulled a small notebook from my pocket and began to write – turning here and there, trying to soak up the essence of a morning and a neighborhood so blessed with sun-lit, El Paso serenity.

It was then that I saw him – the white-haired man who was crossing the street from his yard behind me.

"Who are you?" he demanded. "What do you want?"

At first I did not answer. I finished making a note. Finally, half-turning, I said, "I'm a citizen of this town and I am standing in a public street." I smiled a bit and looked at him more directly, hoping he would appreciate my reply — would take it as an appeal to his basic good sense and back off from his aggressive stance.

But he would have none of it. With his head and shoulders bent at a slight angle he stared at me and came a step closer — a man in his sixties not used, I thought, to verbal sparring matches in the middle of a street.

"What are you writing?" he asked next. It was not a question; it was a command.

Now I have to tell you: I was in no mood to give him satisfaction. If he had bothered to introduce himself and make some kind of courteous inquiry, I would have probably indulged the man: told him right off I was a teacher, that I wrote. I would have humored the guy. But his tight-lipped and blunt interrogation was a patrician's demand that his questions be answered by a lesser mortal: in his mind we were not equals, socially or any other way. By pausing there on the street I had become the accused and he the accusor, judge, and jury.

"Look," I said. "It's a nice day and I don't want to ruin it. . . . If I had thrown a rock through the window of that house over there, or sprawled out on your lawn, then you would have a right to ask me questions. But this is not your neighborhood; you don't own it and —"

"It *is* my neighborhood," he shot back. "I've lived here for 30 years. And I want to know what you're doing."

We were less than a foot apart. He stood peering into my face, his mouth slightly quivering. He was definitely worked up — wanting to know exactly what kind of strange fish he was dealing with.

It was a stalemate. I felt no obligation to satisfy the curiosity of a man who suddenly appears before me and demands: Identify yourself. Justify yourself. Explain yourself.

The man turned on his heel and went over to stand on his lawn. He continued to glare at me. I walked away. I had almost reached my car when it hit me: So that's what a Mexican American, minding his own business, feels when he is stopped by the Border Patrol: that instant rage, flush of anger, stubborn refusal to cooperate when asked to Identify, Justify, Explain simply because he is on an El Paso street and has a brown skin. Sure, I knew the role of the Border Patrol: illegal aliens pour into El Paso and it is the job of the men in green vans to find them. And sure, I understood where the white-haired man was coming from: he felt protective about "his" neighborhood and had seen a guy in old tennis shoes staring at things – a guy who was writing something down. Certainly a suspicious character. Better check him out.

I believe that if I had stayed there much longer – undocumented – the white-haired man would have gone inside his house and called the police. They would have probably answered his call and – to satisfy this worthy and allay his fears – would have asked me to Identify myself. And I think I would have refused to do so – would have bowed up indignantly as I did when accosted by the Concerned Citizen; would have stood there and asked, steadfastly: Why?

I wish now that I had tested my rights – found out if I would have actually been arrested in my favorite city for pausing too long to look at the sky.

Outside Whataburger

At six o'clock on a sensationally pleasant May afternoon I was ready to open the door of the Mesa St. Whataburger. The sky above the mountain was full of exotic, purple-white clouds, the Phillips station across the street was selling regular gas for 92.9, the Mesa traffic was moving as smoothly as fish in a stream. All seemed routine on an after-work Tuesday. Yet as I paused at the door I saw a girl – possibly a college student – sitting outside on a concrete bench, chewing her fingernails. She was sunk within herself, not at all pleased by what was on her mind. And – to be truthful – I was not too pleased with my own state of affairs.

So I didn't move. I stood there as an old and nagging question intruded itself between me and the Whataburger door: Will we ever be satisfied – any of us?

I'm saying that the sun was out, greenery dotted the nearby streets and yards, the peaceful hum-and-flow of traffic dominated a nice-enough time of day in a nice-enough part of town. Yet the girl on the bench was frowning and sawing away at her fingernails as if she were on Death Row; and, a moment before, I had got out of the car feeling that my life was a continual leakage down a bottomless hole.

(I thought: When — if not today, if not *now* — will she or I be content: be happy? Do I know anyone who is?. . . . What, exactly, do we want — in our marriages, jobs, lives? Why such a feeling of incompleteness, of vague desire and vague loss? Why such a feeling of: I can't quite connect with that big elusive thing that I want so much?)

Life was all around me in glorious Living Color, but inwardly I remained a dissatisfied black-and-white. But why? Wienerschnitzel was *there,* and Piggy Bank (premium 96.9) was *there.* Safeway, Thunderbird, 7-Eleven — the whole modern convenience apparatus was intact. So why should the girl look so forlorn? Couldn't one find salvation at Save 'N Gain?

What impoverishes us, I wondered — my hand still poised at the door. What makes us so dissatisfied with what we have and yearn so unrelentingly for we know not what?

I stood there, wanting answers to materialize out of the lambent air and tantalizing sky, then finally went inside.

The young man in the Whataburger cap asked brightly: "May I help you, sir?"

I was tempted to say: "Yes — does modern life really represent progress? Is consumerism a false god? Is less more?"

But I decided against it — settling for a cheeseburger and a vanilla shake.

Montoya Canal

The Montoya canal lies hidden in the Upper Valley like a strip of mercury in a thermometer — invisible to the hurried eye. Yet for those of us who know about it, it is our West Side Suez.

When I walk its banks – past the alfalfa fields and horse tracts on the east, the shaded back yards with globe willows and orange marigolds on the west – I am no longer in El Paso. I am in some lost valley of the gods.

I do not know how the Egyptians viewed their deities, but the Franklin Mountains – rimming the horizon – always look like gods to me: constant, unknowable, timeless. They are not the Rockies with postcard-pleasing snow and pines; they are not the rolling green mounds of the mid-California coast. They are like a granite arm of the earth flung out across the desert toward eternity. They satisfy a human need for permanence and grandeur.

Sometimes I stop – grim joggers passing by me with labored nods – and watch a gray-haired woman in the distance. She stands in the shadows of her Emory Road trees, in her little park of elms. She gives a whistle of sorts and then with an old woman's bare effort she flings several windfall apples across the fence to her horses. A black mare with a glossy coat and the knee action of a Music Hall Rockette moves into the sunlight of the field. As the horse begins to eat an apple a bantam rooster crows nearby, the woman folds her hands into her apron in the elm tree shade, and I move on down the canal – listening to the Inca doves court each other seductively in the greenery of the cottonwoods.

Everydayness

Impressionists constantly paint their bowl of fruit. I could paint, again and again, "Five o'clock at Copia and Pershing Drive."

I stand outside the Circle K eating a Snickers bar and wonder how a city street in Pompeii or Rome would have looked – would have *felt* – at five o'clock on just such a spring afternoon: if it would be any different from this one. What gives a person the sense of a timeless moment – of everydayness – in any century, in any town?

Nothing unusual is going on. Same white-aproned, listless clerk staring out the window inside Circle K. Same pair of school kids passing by, kicking a flashlight battery like soccer players. Same construction worker in his truck with the loose bumper stopping in to buy an after-work quart of Coors. Same girl smiling into the receiver as she leans against the pay telephone. Same cars stopped dutifully at the red light. Same El Paso mountain looming above the trees and roofs of houses.

Yet there it is, suddenly, in the air: the quiet sense of "This is how the world goes, the world goes. . . ." The sun keeps shining down on the earth and someone, paused for a moment, registers the impenetrable strangeness of human dailiness: on a side street in Pompeii or on the corner of Copia and Pershing Drive.

Back Road to Las Cruces

"The aspect of things that are most important for us are hidden because of their simplicity and familiarity. . . . One is unable to notice something because it is always there before one's eyes." So said Ludwig Wittgenstein in "On Certainty."

So said I as I drove to Old Mesilla — seeing ahead of me the jagged Las Cruces mountains rising to meet the summer-noon haze, seeing to the east the reddish Franklins stretched along the horizon like a gigantic statement from the past that had been preserved in granite.

I looked out at the day — just that: the day — and it was there in the summer light as I first saw it as a child: when I was profoundly innocent of adult knowledges, when a single day was the same as life itself, when each morning and afternoon meant possibility and promise.

That kind of day — full of shining surfaces, breezes, pleasing rural sounds — was outside my car window. The beckoning lure of life was in the upward thrust of the mountains, in the midday gleam of irrigation water that ran beside the lettuce and alfalfa fields, in the globe willows standing at the curve of the road and the *whurr* of blackbirds riding on springy salt cedars. The green fields and brown shining earth carried a simple message of renewal, and the sun spread it across each weed and child and dog and rose bush and orange crate. It was free for the asking — which was to say: for the seeing, the knowing, the understanding.

Just before La Union I took a walk down one of the canals, through a pecan orchard, and it was like a park beneath the trees: cool and shadowed and still. I stood there, the sun on my back, the pecan trees gracious in their immobility and deep shade. I took a green pecan from a limb, scraped my fingernail across it, and smelled the rich, tart smell. It was a little thing to do — to smell a pecan — but that is what the road to Old Mesilla is all about.

I drove on, slowly. From time to time I pulled over to the shoulder and let faster cars and trucks go by. I was in no hurry.

At Chambarino I passed an adobe house with washing on the line, a dog sleeping by a door, a boy playing with a stick in the small dirt yard. It was like a Platonic Idea framed in the car window: Boy, Shaded Yard, Smooth Packed Earth, Timeless Summertime.

At La Mesa I got out again, stood beside the car, looked, turned, looked some more, turned again. . . . What a remarkable strip of earth I was standing on. The cotton fields were spread out like gigantic green quilts sewed with yellow thread, and above them the clouds sat on a dish of deep blue. And the sun: it shone and cleansed and illuminated and made perfect every leaf on every tree. It was the land's antiseptic and sustainer, beautifier and companion.

And the mountains: At four o'clock they were solemn intermediaries to distance, to the sky.

And the afternoon silence: it was not something separate but somehow *consisted* of the fields and mountains and sun-washed space. Birds on the wing seemed to know this secret, could thrive on it like an elixir.

In Las Cruces I ordered a hamburger and fries at a Lota'Burger and sat outside at a small table, reading. I ate and I read.

Then I looked up, looked east, and once again I understood why I had eyes, why I drew breath, why I live in the West.

The sky had turned overcast, there had been late afternoon showers, and now, just before sundown, rain was in the saw-tooth Organ Mountains.

My table offered a wide-open view. In front of me was a low line of brick and adobe houses shining in the sun. Beyond them the Organs lay behind a film of sun-lit gauze, and violet-colored cloud islands stretched out in long archipelagoes.

To the left of the main peak a red-yellow-and-green gash — a dramatically flung segment of a rainbow — arched downward into the mountain mist and upward into the overhanging clouds.

The sky was a monumental palette of pastels, and the bright multicolored connective arc an almost metaphysical slash: an otherworldly signature.

A breeze moved the pages of my book, the sky was still an ocean awash with summer blues and grays, and the solid, purplish Organ range loomed like Olympus, taking as its due the moment of sundown grandeur. Shadows lay in the gorges; the mountain tips were tinged with rose and pink: desert cosmetics the gods allowed themselves from time to time.

Suddenly, before I could take the next bite of fries, the rainbow gash extended itself into a gigantic arch from the horizon in the north to the horizon in the south: an all-encompassing Metro-Goldwyn-Mayer extravaganza across and above the sky, framing all in sight: mountains, clouds, rain, sky.

And just as suddenly — behind me — the sun went down, the east turned bruised and faded, became private and withdrawn. Day died in an instant, and the Lota'Burger sign began to put forth its pale attempt at neon glory.

Old 'Catcher in the Rye'

Such a head of gleaming white hair he had, the old Mexican man in the plain cotton shirt. How dramatic it was against his smooth brown face.

He was sitting on a bench at the Sunset Heights park, watching four neighborhood children play park baseball (bat: a girl's loafer;

ball: a green tennis ball). Infinitely indulgent, the old man was pleased by every shout, every hit, every miss, every movement of the dark-haired children who ran through the elm tree shadows around their imaginary bases.

Wearing a Cesar Romero moustache, creased khaki pants, and brightly polished dress boots, he was an alert and dapper spectator. As the girls ran and the tall boy pitched, the old man watched intently, smiling and lightly patting his leg in time with the pleasing rhythms of the moment. (Ah, could there be, for anyone, a more appealing late-afternoon sight: lively young elementary school girls — shoes parked neatly beside a tree — running on shaded, sun-streaked grass in white cotton socks while their blue-socked friend dutifully — tirelessly — pitched to them as though it were his chosen lot in life).

After a while the old man put on a small neat hat and went into the game. He took his place out in field and stood there, patiently. A slim girl in a short green dress waited formally at the plate as the boy pitched the tennis ball. The girl swatted it with the shoe.

The old man, smiling widely behind his moustache, moved toward the fly ball with short, almost mincing steps — with the mock ineptness that adults sometimes use when playing with children — and caught it with both hands cupped together. Then he tossed it with a properly playful straight-armed throw to waving, almost dancing hands as the girl in green dashed like a nimble-jointed marionette around the bases.

. . . I'm saying that this was taking place at seven o'clock on an afternoon-turning-slowly-into-evening, with the El Paso sun streaming in from the west like beneficence from on high; with elm and mulberry trees standing around in a late-afternoon greeness that was as strong and deep and hypnotic as green fire; with the children jumping and laughing and throwing their arms about in the shade as if they were children who would never age, never die. And the air beneath the trees: it was like air-crepe paper, air-ice cream, air-velvet.

After a while the children tired of baseball and as though at a secret signal ran across the park to the swings. The old man, still smiling, followed in their wake. They swung and swung, the white socks of the girls lifting into sun-haloed swarms of gnats, the blue socks of the boy alternating in a lazy counterpoint.

Then, abruptly, they quit the swings, raced to the trees, grabbed their shoes, raced again across the park and down a side street and out of sight. The old man, in his hat and gleaming white hair, limped tolerantly after them.

ON SUNSET ROAD

I.

We live in the Upper Valley near the greenery of the Rio Grande. We thrive near elms and cotton fields. I am content. My half acre on Sunset Road is my private island in the universe bathed by a radiant desert light. It is a place of trees and animals, children and stabilities. As I water the front grass I can watch the cars go by; when I finish I can sit on the porch and watch the cars some more. Time flows evenly in measured, human-sized amounts.

I enjoy simple, sensory pleasures. I walk around drinking tall glasses of iced tea flavored with mint from the yard. I smell beans cooking on the stove. On Saturday mornings I stand with a cup of coffee in the kitchen doorway, thinking, pleasurably, "It's Saturday morning" – and it is as if we have all been rewarded or blessed.

I go outside and stand a while looking at Fellini, our sheep, grazing the back lot. I visit the quail in their cages and the pigeons sitting in the low branches of the elms. I look at their watchful pigeon eyes as they sit there looking back at me, and I know I am glad to be living in my peaceable kingdom.

On late afternoons I watch our black rabbit leave the shrubs of the church rectory across Sunset Road and come down our long plank fence. He is returning home from the Great World before dark – coming back to the security of the other animals after having ventured abroad to St. Matthews during the day. He noses under the fence and goes to his accustomed place under the big elm. He eats his alfalfa pellets, cleans himself a bit, then sprawls out beside his bowl.

Sometimes I go outside after dark to walk around. Neighborhood kids are still riding their bikes along side streets, offering a few last-ditch yells before surrendering to the demands of bedtime. Stars are out. I listen to the night as I sit for a while on an overturned bucket beside the rabbit. He never seems to think I am intruding. He just flattens his ears and stretches out a bit more comfortably. It is a good moment, I think, for both of us.

II.

I get up quickly and dress in work clothes for the day.

Deborah and Byron are still asleep, their mouths slightly open, hands and legs and arms limp over the edges of their beds. I look in at them for a moment, then go to the kitchen. Judy, as usual, has made coffee before she left for work; I pour a cup from the stove and walk outside.

There is a good breeze off the desert rains of the night before, and it is cool against my nylon shirt as I walk about, sipping the coffee. Every now and then the breeze rises, stirring limbs and leaves of the elms.

The June morning is like a big pleasant valley suddenly encountered in the midst of harsh, barren mountains: it breaks the spell of summer days. . . . The cats lie curled in the middle of the driveway, sleeping; Fellini grazes behind the red plank fence; mockingbirds are out over the cotton fields, dipping, rising, circling between the green plants and the gray sky. A tranquility bathes the land.

Byron, age three, wakes up and gives his familiar cry to the quiet, early-morning house: "I-wanna-bottle-choc'late-*miiiiiiilk*!" I bring it to him, and then, an hour or so later, he goes outside in his pajamas to greet the day. He renews his friendships in the yard: first, with his sack swing — a brief ride or two, small bare feet rising above the horizon of cotton field greenery; with the honeysuckle vines beside the front porch — picking off the blooms, pulling them apart, examining their sweet-smelling intimacies; with his dump trucks, his hammer and nails; with the cats, the dogs. He wanders and touches, reacquainting himself with summer-morning pleasures. He sits for a while, digging again in his tunnels, patting his roads.

I watch him and am satisfied. Even though he is not growing up on a ranch or in the back yard of a small hill country town — my own preferred places for children to spend their days — he is nevertheless contentedly at play in good elm tree shade.

After a while we walk over to the Montoya canal. Byron puts his hand in mine and we saunter past the long back yards behind Emory Road where horses and chickens and goats move in the sun, past the cottonwoods growing along the bank, past the big leaning mulberry tree where we stop to stain our mouths and hands with red-and-black berries.

III.

At five o'clock Byron is up from his nap and comes outside in his red tennis shoes and undershorts to play in the irrigation water. The ducks have moved from the cool shaded ditch where they have been swimming and diving since noon and are now nuzzling loudly in the sunny, flooded grass in the middle of the lot. They seem giddy, almost overcome by this sudden wealth of water, and give frantic accordian wheezes of delight.

The turkeys walking nearby go *perk*, the ducks continue to suck noisily in the shallow pool of water-and-light, the huge Rhode Island Reds plod about in the wet grassy strangeness.

At supper I eat with my daughter – just the two of us at the kitchen table. I fix ham sandwiches, and then after peeling a couple of peaches I put sugar on them and serve them in a dish. There are a couple of left-over pancakes from breakfast, and jam, and half a pear, and orange drink.

I look across the table at Deborah and consider her future. I watch her eat the inside of her sandwich – leaving, as usual, the outer edge – and try to imagine her as she will be later on in school, and married, and with children of her own. I sit there, knowing that nothing I can ever do will save her from pain, or disappointments, or the whims of fate. I watch her, sitting in the green curved-back chair, and listen to her shrewd eight-year-old's questions that make chills go up my back: innocence-in-pigtails starting to unfurl her sail, beginning her long drifting out to sea. . . .

I leave the children and their unknown futures – Byron in the bedroom with his coloring book, Deborah still swinging her legs at the table – and go stand for a while in the back lot. It is the Saturday after a week of rain and the scattered patches of grass are a confident, paint-can green in the sandy soil, and the garden tomatoes hang rich-red on their vines. The afternoon thundershowers have moved on and left the clouds and the mountains in a sweeping display of heroic desert sky. In the vacant lot beyond the fence the sun glows like varnish on the new lumber of a half-built house. There is a barnyard mellowness under the nearby trees as the chickens and turkeys and pigeons settle in their cages for the night. High above the Upper Valley a hawk coasts in the currents of a sweetened west wind.

IV.

At eight o'clock I am standing on the concrete porch in the back yard. . . . No, I am *part* of the back yard, part of the dying April afternoon, part of the coming night. I am standing on my familiar back yard stage. The earth is still now, and the scene before me, around me, is focused and intense.

I hear voices from the neighbors' yard beyond the back rock wall – the Dahills' house. There are children there, as always: the yellow-haired Dahills, the black-haired neighborhood boys: Garzas, Acostas, Villareals. My son is there, with Steve, his friend. They are standing by the unseen trampoline, watching as the older Dahill girls finish their gymnast routines. I can see blonde Dahill hair rising rhymically above the wall.

The elms stand in the lots, darkness within their branches. Beneath them, by the garage, the white ducks, bedded down, are babbling to each other in a subdued chorus. At the red fence, near the back lot gate, Fellini stands rubbing her head against a plank.

I leave the porch and go stand at the gate, listening deeper to the yards, the voices, the nearby streets. Small doorway lights are already shining on some of the houses. Down the fence, on her chain, Sniffles the wirehaired terrier stands spraddle-legged. She gives sporadic barking but does not have her heart in it now that the ducks are out of sight.

I listen: Behind me, from the house, comes the high barking of Shadow, my daughter's white Westhighland terrier. I do not turn but I can see her, imagine her at the front screen door, barking at the horse and rider going slowly along the side of Sunset Road. I hear the fading hoofbeats against the gravel.

Fellini finds my resting hand, comes closer. I feel her warm breath against my fingers. We stand there on opposite sides of the fence as frogs begin their chorus from the canal, as the moon rises into the elms.

V.

I get up from the dining room table, take the dishes into the kitchen, pick up the book I had laid aside, and walk into the living room and sit down.

I do not need to move briskly. It is a holiday, and I have plenty of time to do whatever it is I want to do.

The living room windows are open, and the barest hint of a breeze moves through the back yard trees and comes into the house.

Cars pass steadily on Sunset Road, and the sound of their motors and tires makes an unobtrusive ebb and flow.

. . . What I am saying is that the house is at peace with itself and so is the day. So am I. I have eaten, the chair is comfortable, the book is a good one. Three o'clock on this casual afternoon is so pleasant it is as though I have been born just to experience it.

Yet I cannot handle such peacefulness — or rather, I cannot let it alone. I cannot just let it be. I sit, I look toward the door where the outside light is reflecting the front yard grass and fence, framing them — eternalizing them — within the glass, and before long I lay my book aside and am muttering to the air: My God, this is a good moment — too good to let go unnoticed. . . . Now, what can I *do* with it — this timelessness of the universe stilled so neatly here in my living room.

And with the book open on the table and the cars still moving rhythmically by, I ruin, once again, a moment of pleasure. I sit staring, listening, trying to grasp the shape of pleasure — trying to plumb its depths — and it disappears before I can get around to enjoying it. Out of some need to understand and make mine that which is special in life I manhandle bliss until it melts away.

VI.

We have a small old bantam rooster who always gets beaten up when he fights his nemesis, the larger, glossier, younger rooster. The old bantam is no longer effective in a fight because his spurs have curled up like twisted vines. I let him run loose with the sheep and ducks in the back lot after I cooped up the younger rooster with his five bantam hens. The old one likes his freedom and does a lot of muttering and crowing as he walks along beside the chicken-wire pen — sometimes stopping to engage in mock-battle through the fence with the younger rooster who raises his neck feathers and lunges at him from inside the pen.

Yesterday as the old rooster rested in the dust and shade beneath the elms, the male duck mounted — for what must have been the dozenth time of the afternoon — the female duck. The male had chased her out into the heat of the lot, grabbed the short feathers of her head with his beak, then climbed aboard her back.

Out of some private sense of outrage or dismay, the old rooster rushed from his place in the shade, jumped into the air and butted the male duck from behind, knocking him off his perch. The male duck recovered and, bill lowered, turned to chase his critic. But the rooster, muttering and babbling, was already on his way to safety. He barely made good his escape — the bill of the duck just inches from his scaly legs — and flew to the top of the coop. He balanced himself a moment, then reared back and let loose a thin, victorious crow: to the duck, the glossy young rooster beneath him, the whole Upper Valley.

VII.

This summer I have a job to do. I watch 30 elms, large and small, as the elm leaf beetles try to destroy them.

I don't really have to watch them — watch as the beetles eat the leaves, watch as the leaves grow brown and brittle and begin to blow across the yards in the wind — but I do it anyway.

I seem to have no better thing to do with my time than water patches of grass, fix a sagging gate, watch the elms. (And the mountains beyond. And sunlight in the air.)

I am only fit for this kind of thing. Day watching. Tree watching. Watching the two thin-legged, inelegant white turkeys as they walk about in the heat.

And I do not know if it is a summer of sickness or supreme health: sitting in a chair, looking out.

VIII.

I have with me a clipboard, pencil, paper. I am sitting on an old chair beneath five small elms that make a nice canopy of green. It is the long moment after sundown.

I am in the back lot where the garden used to be. Byron is batting dirt clods with a stick. Fellini is nosing through islands of grass that are smooth from her daily grazing. The pigeons walk nearby on another stretch of grass.

. . . Eight-thirty on a July afternoon. All cells are in equilibrium, all pulses are serenely beating. The sheep, my son, the pigeons, the neighbor's quiet yard: We are in the calm eye of creation's daily hurricane.

Thus I can say, in this green-leaf cave: I am totally, consciously alive, and I have no wants. There are no mountains to climb, no metaphysical puzzles to solve. Whatever is absolute is here before my eyes, in the back lot, in this ordinary moment before dark.

Creation is this time-and-place – invisible as the roots of trees, relentless as a dirt clod falling.

IX.

I am on the front porch, watching day begin. The sun is just above the Franklin Mountains. Six-thirty shadows-and-light angle across the yard. It is a tender time.

The first of the joggers passes by – a black-haired young woman with dramatically muscled legs. A gray German shepherd runs beside her. They move so silently they scarcely seem real.

Across the front cotton field comes the faint sound of motors – road graders, perhaps.

Birds are down from the trees. The church parking lot is theirs, cool and shadowed. I look at them in their early-morning briskness and at the neat brick symmetries of the church office building.

I breathe in. The air is still cool.

A rooster crows somewhere in the neighborhood.

Cars begin to pass, their lone drivers sealed behind window glass, headed toward Doniphan, toward I-10, toward town: to begin, once again, the day's business.

Here, on Sunset Road, the day's business is still shapes and colors and sounds: a row of tall cannas along the driveway of the church rectory – like bright-red flags already in place to salute the morning; Mt. Cristo Rey in the distance, well-shaped and -shadowed: a classic mountain-under-morning-clouds; Charlie the cat lying in the marigold bed that is still cool from last night's watering.

. . . The whirring, purposeful, awakening world – its cars and trucks are beginning to speed by now in a steady stream. There is no money to be made sitting on front porches. Every car is an indictment of me in my chair, the whir of its passing telling me: Get to work, get to work.

X.

It has been a hot, difficult Sunday with occasional family disagreements, loud words, angry silences, and now, at 9:30, I stand at the kitchen sink, washing the supper dishes. Byron, in his bed, lies sprawled on the sheet. My wife has taken her second Excedrin and gone to the bedroom to lie down. Deborah is in the living room playing the piano.

The house is cool because of a late afternoon rain; windows are open to the sounds and breezes of the night.

Deborah is playing slow, halting notes of "I've Grown Accustomed to Her Face" as summer-night cars go by and dogs bark in the distance.

Finally she closes the piano top and puts her music away. As she walks toward her room Judy calls out to her, quietly, "I love you," just loud enough for Deborah to hear.

I wash another cereal bowl, slowly.

XI.

I live alone. My children visit me.

Byron, 11, half-runs across the back yard, heading past me toward the gate. He does not run full speed because there is no sense of urgency: He is luxuriating in the knowledge that he is home again for the summer. Yet he runs instead of merely walking because he is in fact here at home after a three months' absence.

He bangs through the gate, scattering ducks and causing Fellini to rise awkwardly in alarm from her bed beneath the chinaberry tree.

"Dad," Byron calls, "what's wrong with the duck?"

"I don't know," I say, and come over to look at one of the females. Her left eye is shut and crusted.

Byron half-turns toward the pigeon and turkey pens – still on the move, still checking out the territory. He is pleased at his discovery of the duck's eye – pleased that he is once again part of the life of the house and yard, noticing things, telling me about them. It is almost as if he has never been away – as if there has been no divorce, as if nothing has changed, as if running across the back lot to his buddy Steve's house is something he has been doing all spring.

I tell Byron we will certainly have to watch that eye for the next few days – purposely saying *we*, wanting the casual sound of it in the air; wanting, like Byron, to establish old routines.

I go back to raking the lawn. After a while I hear voices, then see them, on bicycles, riding out of the Dahill garage. It is a familiar sight: Byron in front, his legs pumping briskly, Steve behind, both heading toward the church parking lot.

As I watch Byron calls out: "Hi, Dad," he says, not slowing, legs still pumping, blond hair shining in the sun.

Just two words, said in passing – unnecessary words, really, because he had left me in the yard not five minutes before.

But they are not a greeting. They are Byron's final affirmation of his return, of his being once more in the long vacant lot beneath the trees. He is comfortable, he is where he wants to be for the summer, and he cannot contain himself. "Hi, Dad," he says, instead of "Look at me: I'm here, I'm home."

THE HILL COUNTRY

Gilmer Street Journal

For twenty years I lived, taught, raised a family in sunny and serene El Paso. It was a good place, and I thrived on its desert air and desert clarities. But my personal Four Horsemen – Death, Divorce, Duty, and Despair – changed the rhythms of my life and thus at 50, a year after my father died, I decided to leave nachos and margaritas and go back home for a while to central Texas: to Kerrville, to the cedars and live oaks of the Hill Country. My mother was there in a nursing home; my daughter was enrolled at the local college; my son, 13, lived nearby in San Antonio with my former wife. The focus of my life, you see, was far to the east of Juarez and the Rio Grande.

What follows is part of a journal I kept as I camped in the old house I was born in and sifted among half-buried memories of my home town.

I.

Incredible: I am in the Heavenly City, driving down Streets of Light. People move along the sidewalks of Water Street and do not see the strangeness of their town, of others' bodies. They think it is only another Monday – this awesome, sun-washed, sun-pierced day. They think they have seen it all before, many times, but of course they have seen none of it. It is as removed from them as Tibet.

. . . The light, the light, it is everywhere – on the sides of people's faces, the fronts of buildings, on the grass and ground and streets, under the blue-domed sky. Women carry packages, ranchers drive pickup trucks, children hold their mother's hand. They pass one another in this exotic, eternal, familiar place, and to them it is only double-stamp day at Super Save.

II.

The yards of the home place have turned into Brazil. Beggar's lice, ragweed, purple thistles, horehound are waist high where only rocks grew before. Hackberries and ligustrums sprawl, interlock, shut out the sun. I get a saw and weed slinger and begin to win back the territory. Then, resting, sweating, I look at the backyard and remember where I sat as a child among the oak tree roots, playing cars. The clothesline ran nearby — holding up my father's newly washed khaki shirts and pants — and the back porch steps were behind me, still splashed with the soapy dish water my mother had just thrown out.

As a six-year-old, I scraped roadways among the roots. I sat in dirt and found contentment there. I enjoyed being underneath the trees, near random tin cans, near rocks, near chickens fluffing themselves in little dirt bathtubs, near the sound of the dogs barking in the neighborhood.

I was small and I lived close to the ground and *lo, the ground was good.* Those big, rough-barked, leisurely curving roots were my friends — as the woodshed was, and the cow lot, and the oat field.

Childhood: that was when my father ran a feed store and my mother was always home. Blueing bottles sat on the back porch window sill, and cane fishing poles lay across the rafters in the garage. Childhood? Why, that meant scabs on elbows and the Avon lady; Sears catalogues and "Lum and Abner" at 6:45; Big Little Books and flowered wallpaper and vacant lots; quilts spread along the creek bank and the smell of burnt tapioca on October nights.

. . . Yes, I fight the weeds of change and decay in the 80's, but the 30's live on.

III.

Backyard, summer day: The locusts are here, of course, rising and falling like Greek choruses within the trees.

After the summer rains the greenery is oceanic — weeds, grasses, shrubbery; hackberries and oaks; they surround me like a sea. Above me, the sun burns down like a friendly lamp fixed forever in the child-like blue of the sky.

Butterflies drift and swim across lantanas like girls picking flowers in some Golden Age.

And in a neighbor's yard, beneath the shade of a pecan tree, a white tablecloth on a clothesline flows back and forth in the breeze: the tolerant metronome of a summer morning.

IV.

Ike Morrow walks knee-deep through weeds and flowers. I am behind him in the morning sun. As we walk along he grins at me, cigarette in the corner of his mouth. We are going fishing. I am seven years old.

That's all there is to it – this fragment of a scene – yet it has come to me and will not let me alone.

I seem to be passing through the heart of a physical sensation instead of the ordinary grasses of Turtle Creek. And Ike Morrow, my father's feed store delivery man: He is not just taking me to catch perch but is leading me mythically onward into some godly pleasure-place among the cypresses.

The sweep of the long, unbroken field below Mrs. Hall's farm – that is part of it. Trees are on the horizon, and fence posts and barns, but in the bottomland where I walk – small, bareheaded under the clear June sky – there is a delicious and intoxicating openness. I could not experience any greater sense of perpetual sky stretching over perpetual, bordering greenery if I were adventuring along a coastal plain in Bangladesh.

I carry a can of worms and a fishing pole and I am headed toward a clump of trees, but I do not know if I ever get there or if I ever fish. I remember only that lyric passage across the flatland: the romance of sweating and breathing and walking as a seven-year-old under the blaze of the sun; the sense of vistas – before me, around, waiting for me; and – in the slouched khaki shoulders of Ike Morrow and the coiling blue smoke of his cigarette – the pleasure of a shared moment, of fishing-pal camaraderie on a Hill Country Saturday morning.

V.

After cutting weeds in the chicken yard I rest a while and look about at the back lot: yes, the home place has seen better days. The rabbit cages are in angles of collapse, their doors rotting in a mulch of hackberry leaves. Rusted rakes lean against the tool shed wall. In the cow lot weeds grow taller than my head. Everywhere I look there are old lanterns, paint cans, coffee pots, rolls of wire.

Yet the debris and neglect of recent years do not matter because the chicken yard this morning is a pageant in the sun. Ragweed towers elegantly beside the garden fence. The young pecan trees smell clean and bitter-fresh. Flies, wasps, bees sashay around in mid-morning odysseys.

In the hen house a White Leghorn walks about within the shining rust of the screen wire. Her smooth white tail moves like a shark fin in the depths of the hen house light-and-dark: now gleaming, now gray. She is serene, effortless, perfect.

I sit sweating in the bright ten o'clock sun until I understand what the morning is all about. Everything in the chicken yard seems perfect simply because the human act of seeing is perfect. To see is to complete a perfect act.

The blade of Johnson grass against the cinder block wall, bending and tapering – is perfect. The red-brick chimney rising above the roof of the house, with the blue of the sky and the green of the trees behind and beside it – is perfect. The sun – shining, creating shadows and depths – makes each leaf, each flash of bird wings, each weathered board perfect. The life-force within a hummingbird, poised at the mouth of the cow lot faucet, is perfect.

The morning gives me intense visual delight simply because I am aware of what I am seeing at the moment I am seeing it. I see each object clearly, as if through a magnifying glass, and almost become the thing perceived, almost participate in its private being.

In the chicken yard everything dramatically *is* – every edge of board, every bug on the ground, every stem of every weed and plant. Ordinary back yard sights and surfaces exist superbly in the sun and I see them acutely – dish pan, broken flower pot, purple verbena – radiant with the presence of themselves.

Byron and I

We have fished the jetties at Port Aransas, looked across Llano country from Enchanted Rock, drifted down the Frio River beneath the cypresses at Garner Park, and now

— well, we are just going to drive out from Kerrville to the west pasture of the family ranch place and camp for a while. My son Byron and I and a dog.

We load the Chevy hatchback with boxes and quilts and the compact, two-man nylon tent from K-Mart that almost fits in your pocket and we ease out into the four o'clock June heat. Byron's solemn-eyed dog Duchess is curled on the front seat by the air conditioner. Byron lies in back among the boxes. He is reading a comic, his legs angled upward over the styrofoam cooler. I gaze at the passing fields, then look once more into the rearview mirror at Byron's toes: they are pieces of sculpture, those white, narrow feet and ankles, those pleasantly curved bones, that shining flesh. I drive on, enjoying the countryside, but I keep glancing at the symmetry of my son's feet freed from shoes and upraised in the air. There is nothing exotic about them; they are just ordinary, human boy-bones suddenly delicate, suddenly strange, like white, undersea flowers that have joined together in the hatchback light.

We turn off the Mountain Home highway and drive upward toward the ranchland plateau. Some of the pastures are unfenced, and sheep lie in the road in the shade of trees. I drive slowly around them, then see it again: the Oehler home. It is a ranch house I have admired — coveted — for 30 years, an unpretentious rock house set in a clump of oaks. There is a garden in back, a neat wire fence, a rock garage, a front porch with an arched entrance. Sitting there, in just the right blend of shadow and sun, just off the road enough, with just the right amount of yard, it has always represented to me the beauty one strives for, the peace one wants, the home in the country at the end of the day.

At the gate to our property Byron gets out, starts undoing the chains around the post. As he stands there, trying to make the worn key fit the lock, I find it a pleasure to look at him: my son, 12, almost 13, in his worn blue tank top, blue shorts, Nike tennis shoes without socks. I like seeing him do that — open the gate finally and swing it forward, then stand there beside it in the rising bit of dust, waiting for me to drive on through: my son the apprentice guitar player, the seventh-grade drummer, the lover of loud "Rush" rock songs; my son, his Prince Valiant-style blond hair shining in the sun, his mother's features nicely chromosomed in his face.

The road to the hunting cabin is rocky, almost impassable; we ease along it for half a mile and stop at the ridge that overlooks the small valley to the south. It is a classic amphitheater of grassland, with tree-lined knolls on both sides. We get out and stand in the country stillness. A buzzard is circling high overhead. A mockingbird is holding a private carnival in the top of a post oak. Thistles – hundreds of them below the ridge – wobble their purple heads like Martian shock troops that have descended onto the ranchland in the heat. They are tall, slender, and spiny, but their heads – all slightly turned, all at angles – are demure, beautifully round, as if they are wearing brilliant Martian afros.

We select a level spot to set up the tent – facing it south to catch the breeze. Byron assembles it in a professional manner – tightening a rope here, realigning another there – and soon it becomes a bright-orange envelope staked neatly to the earth. He calls Duchess; they race down into the valley. Beer in hand, I watch them a long while, the way you look at a painting.

They return, and Byron and I begin to horse around with Country Talk. "Hey-boy," I say, "you better find me some cedar bark if we gonna have any vittles tonight." Byron says, "Shore, Paw," and races Duchess across the rocks and needle grass to a big cedar tree. He returns with enough long strips to start the campfire. Then he gets his BB gun and roams around shooting at gourds, dragon flies, scorpions under rocks. He stops a moment and says in a mock gunfighter's drawl, "Wal, cow patty, we meet again," and fires four quick ones into the soggy pile.

"Reckon we better clear the land, Paw?" Byron calls out. He gets a gray, weather-smooth limb from a brush pile and begins chopping down the tall mullein plants along the ridge with loud, kung fu *haaaaaas.* He stops, stares at a formidable-looking plant in front of him.

"He just called you a bad name, Paw."

"A bad name!" I say in my own Yosemite Sam voice. "Well, *git* him, Son!"

Byron swings his stick viciously, toppling the badmouthing plant.

By seven o'clock the breeze is sweeping the ridge. I sit on a canvas camp stool in the shade of the car, reading. Byron returns from another round of exploration with Duchess. He drinks a root beer from the cooler; his fuel-injected belches rock the air.

Time passes and the ridge is quiet. A plane drones by. Then, from inside the hatchback: "Dad, what does *girth* mean?" A bit later: "What does *bauble* mean?" Welcome questions to me. They mean that Byron has made himself a place among the quilts and is reading his library book. *The Black Cauldron* is a sword-clasher full of huntsmen and enchantresses and characters with alphabet-stew names like Orgoch, Lluagor, and Fflewddur, but I am pleased that he has picked it up — I had tossed it into the car next to his BB gun in order to tempt him — and that he cares to ask about words. This has been a historic summer: he has cut back on watching television and discovered reading. He has made his way through all seven books of *The Chronicles of Narnia* by C.S. Lewis. He likes the Olden Days.

Shadows stretch; mourning doves begin to call. We find a likely place for a fire in the rocky crack that runs down the ridge to the valley floor. It has a good draft, no grass around it to catch fire. We carry big rocks, shape a fireplace, spread other large rocks around it to make, roughly, a surrounding table. We build our fire.

I bring the hamburger meat and the pan from the car and Byron starts cooking. He puts a strip of bacon on top of each mound of meat. I sit nearby on a rock stool I have fashioned, savoring the smoke smell, the hamburger smell.

I go up to the car, get another beer, look about. Streams of eight o'clock light touch, wash, surround the tree trunks as they stand in the pasture. It is an intimate scene — as if the trees had finally taken off their clothes at sundown and were privately, chastely bathing up to their knees in a sea of reddish, primordial light flowing in from the west.

We eat our hamburgers and small cherry tomatoes and Fig Newtons. We put on more wood to keep the fire going. At nine o'clock the first pale star appears above the ridge. Katydids begin to pulse in the trees. We listen: sure enough, an owl is in the trees to the east of us. We listen again; another owl answers, farther away.

I slap at mosquitoes and Byron pokes at the fire.

"Tell me again about the ranch," he says — it's dark now — "when you came out as a boy. What did you like best?"

He is talking about the ranch house itself when my grandparents lived there — when it was still a working ranch, a lifestyle, when there were sheep and goats and horses and mules and hired hands and gardens and grain fields and homemade ice cream and large family gatherings. He has missed out on all that. My grandfather had already died when Byron was born; he barely remembers my grandmother as a shrunken old woman in a Kerrville nursing home who gave him Juicy Fruit gum and orange slices. He has looked at all the family albums, heard anecdotes about relatives he has never seen, stayed in the ranch house before it was sold — has sat in the chairs that Gram and Grandpa sat in, has eaten at their dining table — but it is as though The Ranch was a half-century-long parade that he just barely missed seeing. It's as if there had once been a family mirror that — if he could have looked into it — would have shown him the fullest possible picture of himself, but now it is broken.

So I talk awhile about what the ranch was like when I was growing up. I tell him that those barren, rocky fields he and I once walked across, full of horehound, had been corn fields with stalks higher than Grandpa's head. I tell him about Gram in her bonnet making lye soap in the wash pot on summer mornings — and the clean smell of the laundry flapping on the line. My best moment to remember? Well, I tell him, maybe it was waking up in the feather bed in the west room, next to the garden, and hearing Grandpa's boots clumping and scuffing on the sidewalk and then the gate creaking and slamming as he went into the back lot, and the sound of the windmill turning — slowly, easily, as if it would keep on turning forever — and the lambs bleating up by the water tank, and Gram, with her hair in braids across her head and wearing her apron, coming across the linoleum floor in her worn canvas shoes, saying goodmorning and asking if I was ready for breakfast, and my saying yes, and stretching and turning beneath the covers, and then lying there, feeling perfect, smelling the biscuits through the bedroom door.

We let the fire die down after a while and look up at the stars. It has been a long time since either of us has seen the sky at night in such an isolated country place. We stare, properly impressed — wanting to talk about what we see but knowing there isn't really much that can be said. *The stars.* They're there . . . my God, they're there. Byron knows about starlight — that what we think we see is really light that started out billions or trillions or quadrillions of years ago. . . . We find the Big Dipper.

Duchess begins barking far off in the valley, and we suddenly remember her. Byron first calls her, then he gives a whistle — an astonishingly shrill, penetrating two-fingered blast of sound I didn't know he had mastered — and we hear her coming. We decide she had found an armadillo.

"Want to go up to Harper?" I ask. It is nearly ten o'clock. We ought to go crawl into our orange-colored cocoon for the night. But the question was suddenly in my head.

"Sure," Byron says, "maybe we can get a soda." My son-the-camper: not yet an ascetic, not yet willing to survive on grubs and berries.

We drive back through the gate and up the farm-to-market road toward Harper. Windows down, we move leisurely past the fields of corn, past the lanes and mailboxes — Goff, Klaehn, Wienecke, Kothmann. Byron has his arm out the window, letting it ride the air currents, ready to nab one of the grasshoppers that flare into the headlights. Duchess has her head in his lap.

It is a moment of pleasing darkness on a smooth back road, the wind so cool now it is almost cold. We are going up to Harper, population 383, just sort of *because,* I guess: because it is nearby so why not; because Byron hasn't been there very often; because we are camped out in a pasture under a sky incredibly loaded with the rest of the universe and suddenly I have an impulse to touch base with the human community — any human community; because the drive up that particular road has always been pleasant.

The road curves, eases down into dips at dry branch crossings. Deer graze along the roadside, their heads, eyes, bodies turned slightly toward our lights. Ranch houses sit unseen in an oak flat or on a cedar-cleared rise of ground. Sumacs grow like lush hedges in a park along the fencelines.

We drive slowly into town and down the single main street, looking at the dark stretch of houses and stores. . . . There is the Modern Market where my uncle still buys groceries when he comes in from the ranch. There is Harper High. There is the street to the cemetery where my grandparents and father are buried. There is the pecan grove by the creek where we used to have family reunions when I was a child. We turn off the highway and look into the quiet houses.

"What do you think it's like," I ask Byron, "living here every day in a small town like this – no movies, but a windmill in your backyard?" He doesn't know what to think about it – about living here. Neither do I.

On the highway I stop at the Longhorn Inn cafe. Five or six people are seated at a table, talking casually in the weak overhead light. The ceiling fan is barely turning. A woman gets up from the table; I buy two Sprites from her. Byron and I drive on back toward our pasture, drinking from the cold cans, satisfied with our little trip. Byron belches hugely into the night air.

The Creek

On an August afternoon I drove down from my Uncle Doug's ranch house to the creek. At four o'clock the cedar hills were beginning to cool, and cows drifted through the pastures among the agarita bushes and prickly pear. I parked at the shallow crossing and stood outside the car beneath the sycamores. Afternoon shadows stretched along the bank and minnows idled in little pools. In the places where the creek was dry the rocks were white and shining in the sun. Small frogs sat in the shallow part of the creek in the mud and shade.

The creek. That's how I had always thought of it, but I might as well have called it the holy place.

A lamb bleated in a pasture, a dove called in an oak, slim green fish swam in the deeper holes of water. A breeze moved through the leaves of the sycamores, and on the ground the shadows of the leaves shifted, separated, blended again — slowly, rhythmically. The white rocks gleamed under the sun.

The afternoon was perfect, the creek was perfect; the only thing missing was words.

We were together, this place and I, in an intimacy, but we could not speak. I could see the Spanish oak but I could not tell it how satisfying I found its leaves. I could hear the nearby, unseen bird, but there was no sense I could make of its intricate call, given over and over — like a message to be decoded.

I stood there, with the tree leaves washing together in the sun and lambs bleating and yucca plants rising like signposts of eternity in the hillside shade, and I wanted to say something, regardless of how foolish it might be — perhaps "Reveal to me, O Creek, your essence, your mystery." I wanted to say hello to the universe — there, in that familiar place — and I wanted some kind of answer: a sign, a voice from a burning bush.

But the rocks kept on lying in the sun and the cows moved steadily through the prickly pear. I got into the car and drove back up the road to my uncle's house. We talked a while on the front steps about wool prices and the screw worms in his sheep.

Selfless Glories

I sat in the back of an auditorium, listening to my son's high school symphonic band play "Festival" and "Carioca" and "Blues in Beguine." I watched fifty instruments move through intricate patterns like a flock of performing birds as the band director — steady, subdued, godlike — guided them along: now encouraging the French horns, now cautioning the basses, now agreeing with the suddenly vibrant saxophones.

I marveled at it: such a sense of joint purpose as the band members sat on stage in their blue uniforms and white shoes: such a feeling of working together in order to fashion melodic excellence for assembled parents and friends. They were proud to be sons and daughters who could be so disciplined and drilled, so capable of producing such smooth, rhythmic sounds.

The drummers stood poised, waiting for their cue; the trombonists kept their horns erect on their knees. As the baritone soared and the clarinets shrilled and the trumpets attacked their notes cleanly, the band members knew – were secretly pleased at knowing – that Music was a bigger thing than they were. In that long, rather beautiful, rather unreal moment on stage, they were willing to seek out the selfless glories that come with group efforts and group effects.

In Bulverde

Strange name but familiar territory: Bulverde. It is not on my map, but I can get to its trees in half an hour anyway.

I live in San Antonio now, and the streets sour me. And the malls. And I-10 and 410 and I-35 and I-37 and the whole metropolitan swarm.

I discovered Bulverde by accident, up the highway toward Blanco, and found that it is hill country enough when I need it.

I went there on a Saturday afternoon, sick with self. The sky was overcast, and the first fall breeze was blowing through the pastures. I wandered among live oaks and mesquites while a Spanish billygoat watched me from his thicket: alert, and properly suspicious. Then, like a boat, a swimmer, a muscle-bound athlete he moved on out into the clearing, legs and head bobbing in slow counter-motions. He stopped and looked at me, snorted, moved on, still alert within his goat-domain.

I idled through weeds and yellow flowers, past cactus clumps with their dusty, purple bulbs. I stopped and wiped a patch of white lichen from a cactus lobe, rubbed it between my fingers, and, like the first astonished Indian, watched it turn blood red. I picked up acorns – green and shining and symmetrical – and simply held them, looked at them – wondering once again why there was no valued human use for such pleasing things.

I stood within a curiously formal circle of oaks, acorns in my hand, and it was as though I had come to another Stonehenge for private, redemptive rites. I stood a long while, and the trees and the wind bathed me as tenderly as a nurse washes a burn victim.

The hill country, I thought: Everyone should just leave it alone, undisturbed, unurbanized. Let it be as the cows know it, and the quail. Let it catch the sun on its rocky ledges, its yuccas and agaritas, instead of on condominiums. Let it alone — and let it heal you, let it bring you peace.

I walked some more and came across a small family graveyard on a shaded rise of ground. "March 17, 1895-March 20, 1895," one faded gravestone read. Walter Bartels didn't make it to the twentieth century, didn't make it a week . . . didn't do much of anything in Bulverde except get named . . . and there he was, lying forever beneath his slab scattered with leaves, while the rest of us went on doing the living he didn't do.

I took out a book of D.H. Lawrence essays from my back pocket and read a while beside the graveyard fence — with fall in the air, and cedar smells, and the summer dying. I read as slow, sleek horses moved about in the pasture: horses with such bellies of grace, such tails of ease, such brown skin shining, that you wonder that they have never been worshiped as gods.

And as I sat there I found I had done it again: I had taken a day that had been sterile, behind glass, and had made it intimate, and personal, and real.

. . . It has never failed me: Give me a stretch of land and let me walk on it a while, give me weeds and wind, sunlight on a fence — give me *country* — and I am home. I could die in the country or live in the country — somehow it would not matter which I did, for I would be all right in either case. Trees, sumac bushes, oat fields always give me back life whenever I seem to lose it. They give me a beauty and truth I can find nowhere else. . . . I had rather sit with a clump of needle grass than with the Pope.

So I go to Bulverde and walk the side roads. They affirm my right to be me, and I breathe air that is properly mine. I am alive, and in the right place. And as the good D.H. said, "To be perfectly alive is to be immortal."

Survival

It was a humid San Antonio morning in June — and I was in no mood for it. School was out for the summer and I was tired — tired of teaching, tired of still another busted love affair, tired of loneliness, tired of life.

So I drove to Kerrville, bought a bottle of Tom Moore, and went to the old home place. No one lived on it any more but its porch was still there, and the yard trees. I got a plastic cup, some ice, and poured the whiskey. I sat on the porch, picked up *The Grapes of Wrath* from my paper sack of books, and began to read at random. I read once again about Granma dying in the night with Ma Joad holding her until the family had crossed the desert into California; about Ma sitting there afterwards on the running board of the truck, talking to Tom — exhausted but satisfied because "the fam'ly" had made it across: had made for themselves the chance to find work and a new life.

I read a while, then looked up. A wind was in the tall yard trees, pickups drove past with dogs sitting soberly in back, mesquites in the field across the street were leafed and green.

Patience, I thought; and stubborness for the long haul: that's what it takes to survive. . . . Good moments and good places will always reappear, like rain — like mesquites blooming, like dogs riding in pickup trucks. Life always washes over you again, no matter what.

Sitting there, I relaxed. Despite the fact that the Kerrville house had no "fam'ly" any more and was settling into decay, I did not feel like Muley Graves — like a "graveyard ghost." With Tom Moore in a cup and Steinbeck in my hand, I knew I was in good shape.

The Good Life

Outside Bulverde people are moving onto the land. Call it farmland, ranchland, pastureland — it's up for sale. A man from San Antonio can buy several acres and live where the air is clean. You can understand how good it is for them on a weekend: the man works with an electric drill by the side of the trailer; his young wife holds a wooden panel for him. Their daughter, 11 or so, comes out of the trailer and does a little pirouette in the shade of a nearby oak. The son, 13, walks toward a bale of hay that has a homemade target wired to it. He has on his black gunfighter's hat and a kind of crossbow in his hand. The family dog lies under the tree, watching.

Homesteaders. They come with horses — some of them. They move a simple frame house and put it up on blocks in an open space beside the road behind sunflowers and hackberry trees. It's their country place in the making. They are going to rock it, and have flowers in the yard, and in another six months it won't look as if Okies have moved in.

Sweet Bulverde air, open space, hill country. People still searching for the good life.

At the Hill Country Camp Meeting

A Baptist ranch family stands together in the congregation: an almost-pretty, soft-faced, large-headed boy — Orson Welles in miniature; his pretty, soft-faced young mother with rich, brown hair down across her shoulders; and the young husband, taut-featured and attentive to both wife and son. They stand in prayer and they stand in song among the other earnest Protestant backs and shoulders. They are Hill Country Gothic except in one detail: the husband not only circles his sun-burned arm around his wife's waist but rests his large-knuckled hand far down her hip — yea, on her very bun.

The young husband sings in the customary way of camp-meeting males — a tuneless, manful croon. His lips barely move — as if the hymn is written in Chinese and he is going through a weekly pantomine. If his voice could be heard on a hidden microphone it would be the tremulous voice of a shipwrecked survivor making last, faint noises of distress. His wife, however — surely a choir member at her own church on other Sundays – puts forth her strong, steady alto on all three verses of "When the Roll is Called Up Yonder." Young Orson, of course, sings not, hears not. He stands at his mother's side, light years away, engrossed in a scab.

The wife, holding her large black Bible, and her narrow-faced husband – with Paul Newman eyes so blue-white they are like chipped marble – are the only flesh-against-flesh couple in the tabernacle. They stand, a decorous and formal pair, blessed by offspring and sanctioned by their peers: sensual as goats.

The Corner of the Yard

On a Saturday noon in early November I stopped at Welfare – not actually a town but just a country store next to someone's abandoned old house. I got out of my car and stood a while in the deep roadside shade.

The house, set back in the yard behind moss-blighted oaks and hackberries, probably looked the same as it had looked 15 or 20 years before when the last old-timer gave up the ghost or was carried off to a nursing home. Cannas still grew in the side yard; a lilac bush bloomed purple. Unbroken dining room windows still faced the farm-to-market road to Boerne.

I watched the front yard – an elegantly quiet, sun-and-shadowed yard. There were slender stripes of pale November sun – almost not sun, almost not yellow, just thin yellowish light – alternating with mild brown shadows: midday partners beneath the trees. But it was the corner that finally held me, that became personal: the corner where the wire fence joined a cedar post, with the yard sloping a little there and the outside undergrowth making a kind of hedge. It called forth every private place I had ever known. . . . Just that: the corner of a shaded yard; private as Sunday, private as death, strangely beautiful and complete, as if it were there that life was born.

I knew I had looked before at such a yard. It was my childhood I was seeing; I was once again three, or five, or eight years old, looking at childhood strips of shadow-and-light.

The yard seemed to be the essence of everything that had haunted me. It was every house by the side of the road, every fence at the end of a lane, every landscape seen in early fall. It was sad, and elegaic, and it was home. What had my life been but a continual seeking of that original corner of that original yard — with the sun out, and bees gathering in the hackberries, and the sightless faces of windows looking toward the west?

Beside the Band Hall

After the football game I wait beside the band hall for my son to come out. I am a parent waiting in the dark among other quietly waiting parents. We have waited before; we do not mind. Our children are in The Band, and thus we wait, indulgently. The Band is not God, but it is Good.

We had watched again this night as we had watched nights before: the band members filing out of the stands just before the half: purposefully, efficiently, at a cadence: a disciplined and ready group, forming unobtrusively at the far end of the field, a quietly moving line of blue-and-white. And then the first half was over and the band was on the field: drums and horns and uniformed bodies. It was, quite simply, an emotional time: a moment of sound-and-glory. Twirlers twirled, flag bearers whipped and whirled their flags, bandsmen marched briskly though their intricate maneuvers. It was stirring, and professional, and it was performed by our children, smartly in step, before thousands.

So afterwards we wait: proud parents, all. We wait as our children put up their instruments, change from their sweaty uniforms, wind down. We hear them inside the building and see them through the windows.

I find myself wanting to say, as if speaking to some citizens' group: If you had a son, growing up, going to high school, where would you like him to be at ten o'clock on a Friday night?. . . . Me? I'll keep my spot right here at the side of the band hall, in the shadows, waiting for my son to come out: Byron, the drummer, the member of the band. His hair wet, plastered to one side, he will not have much to say — indeed, will shrug and suggest that the half-time show was not up to par: people out of step, lines crooked. . . . We parents know better. We listen, and smile inside.

Clayton Weatherby

I think frequently of a man I've never met. His name is Clayton Weatherby and he lives alone in a house across the valley from my Uncle Doug. I keep forgetting that he must be in his sixties now. In my mind he is still that strange young man who turned his back on people — who returned from World War II and lived with his parents until they died and then just kept on ranching and living alone among the cedar and Spanish oak trees of central Texas.

I have sat out on my uncle's porch on summer nights and looked across the darkness of the valley to the Weatherby house. Sometimes I can see a light, sometimes not. And if I hear the rattle of a pickup and see dim headlights bouncing along Indian Creek Road, my uncle will say, "Well, I guess ole Clayton has been in to see his woman."

If I were around him a while maybe I could learn something useful — how to do without city lights, city distractions, how to back off from the world. But I couldn't live alone for thirty years: Get up each morning and stay there on the ranch during the day doing whatever needs to be done and then after night begins to settle — whippoorwills calling from the pastures and the bullbats gliding though the pens — sit alone on the porch for a while listening to the katydids before going inside to bed. I could do that for a week at a time but not for thirty years. It wouldn't be enough for me — a trip now and then for groceries or to see a widow.

I think of Clayton's light across the valley. I think of him sitting there on his porch — looking over to the distant light of other homes — and I wonder if he is content with being who he is and what he is doing, or if he is simply another cripple unable to endure the stress of human give-and-take. I wonder if he is serene or nearly mad with loneliness. I wonder if he is aware that people wonder about him.

Chances are I will never talk to Clayton Weatherby – never sit with him in the early evening – and that is perhaps a good thing. He will remain an unsolved, and thus unforgettable, mystery: another dim porch light burning enigmatically in the unending human dark.

Stage Set

After the movie was over I edged along slowly with the crowd and out the front doors of the Arcadia Theater in Kerrville. It was night. People got into their cars; motors started; car lights were turned on. Laughter rose from the shadows of the parking lot trees.

That was all: just an after-the-movie moment on a familiar hometown street. But as I walked down the sidewalk toward my car, Water Street – that is, the buildings along it, the warm summer-night air, the couples outlined by headlights – was not just Water Street. It was a stage set and I was the audience.

I stopped beside my car, got out the key, looked around, and said the words out loud, "I love life." Then I leaned against the top of the car. I stayed there a while, unable to move: I could not just get in and drive away – not with the stage set of Water Street still in front of me and the trees of the parking lot still outlined against the sky.

. . . Yes, of course: One loves life. And movies, and small-town streets at night, and couples walking hand-in-hand, and car doors slamming. One loves the humanly familiar and humanly satisfying.

And that's what made the moment so hard to handle. I could say, in a surge of emotion: "It's good . . . God, it's good: each cricket beside the parking meter, each letter of the sign above the jeweler's door. Life's good because it *exists.* It's all we know and it satisfies us and we never want it to end."

But the shadow-thought was right there too: the knowledge that nothing ever lasts, neither cricket nor car nor the human consciousness that is aware of them both.

I stood there, thinking: What is ever real, if this is not? If not this moment beside the Schreiner Dry Goods Store on a summer night in July, if not that other moment inside the Arcadia Theater — when?

"I love life": I had smiled and smelled the night air and felt the sudden need to affirm, to embrace it all. But the longer I looked at Water Street, the more the stage set dissolved. The play was over for another night; the streets were empty.

A Private, Green Place

It was October in central Texas and the rains had come. After school each day I drove north out of San Antonio into the cedar hills — away from a job, a city, a personal life that did not satisfy me.

The drivers of sleek-bodied cars and trucks that passed me on the highway were going to the comforts of suburban homes among the oaks. I was simply following a deep instinct to survive: to get out, get away into the countryside.

Bulverde was just a crossroads community, but it was my chosen place among the fields and hills: a place to park my car and breathe in the rain-washed October air. I walked along the subdivision roads, and the white roadside rocks — bits of caliche left in random mounds by a developer's bulldozer — sat in the dirt like old friends of mine (— at 53 one found friends in unlikely places).

One afternoon I walked out into a field of scrubby cedars — into a private, green place. I sought them out — the cedars — for a kind of healing, though I did not understand the force I was obeying.

If the cedars and I could have talked, I would have said — several things. I would have said, You break my loneliness and I thank you for it. You bring me the familiar comfort of enduring, stable, growing things. You bring me *me* in a time of mid-life despair.

I took the limb of a cedar bush and slowly stripped away some of the green needles. I brought them close, smelled them, let them drop. Then, as I looked down, I idly pushed against the soft, muddy ground with my shoe. I stepped back and stared a long while at the imprint in the mud as if it might somehow reveal to me – there in that congregation of cedars – a truth, an understanding, the face of God.

That was part of the long despair: wanting to break through to new dimensions, to plumb new depths – and always having nothing happen. I wanted to do a thing that mattered as I walked through the days. I wanted to tell of the quiet, eternal hum of the Bulverde earth. Yet often I simply stood in a muddy clearing smelling cedar limbs as the sun went down.

My Uncle Mitch

Every kid should have an Uncle Mitch. He was what kids and uncles are all about.

Mitch met my Aunt Geneva – Neva, to the family – when he was working as an oil field roustabout in the early 1930's. My aunt was teaching school in Texon, a little West Texas oil field town. For two years she was undecided about Mitch – there were other beaus, other plans and possibilities. For two years they rattled along dusty oil field roads into San Angelo, down across the border to night spots in Villa Acuña. Then she decided yes – and it was a twosome, a team, for life. (I've looked at them dozens of times in the family photograph albums: Neva, the heart-faced beauty with plucked, arched eyebrows – a dead-ringer for Clara Bow, the "It Girl" of the silent movies; and Mitch, the big, blond, smiling Irishman: the man's man).

Soon after they married Mitch was transferred by Plymouth Oil to Sinton, 30 miles from Corpus Christi on the South Texas coast. They lived in a little white frame house with crushed shell in the driveway and oleanders in the yard. They had no children those first years of marriage and it seemed as if they never would. They had, instead, each other — and Jerry, their airdale. (Jerry: they stood him on his hind legs between them in the black-and-white Kodak photographs: Jerry at the beach, Jerry in the front seat of their old Plymouth: Jerry, grinning, between the handsome young Mitch and the pretty young Geneva: Jerry, their substitute child.)

Then, as it happens, after years had passed, a child was born: beautiful, by all accounts, as his parents were beautiful: Michael Lanier, who was born on a cold January morning and who lived six days.

Mitch loved to fish, and so did I. When I would visit them he would load Dr. Peppers in his huge ice chest and get his tackle box with the two levered trays filled with hooks and lures and lines and weights, and we would take off for Port Aransas. We would walk out on the pier and Mitch would establish our spot. I would stay with him all day without complaint, casting and casting, as if nothing else mattered in the world except watching my cork on the blue-gray water and listening to the swells as they lapped beneath us against the pilings.

I think that big men, when they are patient and kind, are the kindest and most patient men of all. Mitch never lost his temper at some miscue of mine. If I accidentally kicked our sack of bait shrimp off the pier, Mitch would laugh and lumber off down the pier in his old Stetson and khakis to buy some more.

Since he had no son to show things to, Mitch showed them to me. He took pleasure in my pleasure instead of his own. He showed me how to fix my line for salt water fishing (the weight is placed lower, the triple hook is up above); how to peel shrimp and what part to use as bait (not the head); how to cast and when to reel in; how to clean fish and gut them; what to do when I caught a scavenger catfish and it jabbed me with its toxic gaff as I took if off the hook (put some of Mitch's wet tobacco chew on it).

Mitch talked to me in the same easy way he talked to everyone else: the filling station attendant in Sinton, the grocery store clerk, the cafe owner. He liked people and liked talking to them; he liked to tell jokes and anecdotes and stories. He liked to explain how things worked. When he would take me out with him to the oil leases, he would point with his cigar out the window at a new oil well pumping in a clearing of mesquites and then spend the next fifteen minutes telling me how deep it was, who drilled it, how many barrels a day it would bring. He never talked down to me just because I was eleven years old.

Sure, I didn't stay a kid forever and he didn't always have to sell milk and eggs on weekends to supplement his income. He began to get into the oil business a bit himself and gravitated toward big cars, a big house, big money. And sure, life caught up with him: cataracts, eye surgery, and finally cancer (I would call his hospital room and talk to his nurse and would listen over 600 miles of long distance as he screamed with the pain that morphine no longer eased).

My aunt lives alone now in a two-story house with her 12 dogs, television set, and reclining chair. All the big money is long gone. During my last visit to see her she said she still wasn't ready to "go," but if she did it was all right. She had her cemetery plot paid for right next to Mitch.

Courthouse Square

I am seated on a bench, looking across a wide stretch of courthouse lawn. The morning is still cool under the pecan trees with limbs arching down.

It is Kerrville in the summertime, but the moment is so slow-paced and timeless – somehow so *Southern* in mood even though there is no statue of a Confederate soldier – it could be any place: a campus in Austin, a city park in Birmingham. It is a tranquil place of greenery just to the side of the casually beginning workaday world. Trucks along the nearby streets provide a constant background drone – changing gears, slowing down, gradually accelerating, moving heavily on. And at a construction site across the way an engine throbs and pumps, interrupted now and then by a jackhammer: brief, insistent, staccato. They are curiously satisfying, these steady urban sounds. They provide a kind of Motor Music at a nice remove from the pecan trees, oaks, and magnolias – from the fountain with its shining jets of water splashing into a pool.

A man in a coat and tie comes out the front courthouse doors, legal folders in hand, and says goodmorning to the workman sweeping his way down the shadowed sidewalk. In the distance pedestrians wait patiently for the red light to change. A young mother passes by pushing a baby carriage — her heels clicking briskly, her hair bouncing. Sparrows race each other in quick bursts beneath the trees. The jets of water fall rhythmically through the sunlit air.

. . . I assume that as a child I experienced just such a place — a stretch of public grounds with morning shade spread like blankets across the grass. If I could have put my feelings into words, I might have said — looking at a flagpole and its flag, a granite historical marker, a smooth solid building: "Things in life are being taken care of." . . . Or maybe: "People and life are good." That first resting place in the center of town would have given me important information: that the world was basically a positive place, not negative; that I liked it, trusted it. Perhaps it was when I was ten years old, waiting outside a pleasant old building in San Antonio while a grandparent was inside talking to a doctor about eye surgery. Let's say I was waiting under trees with the muted big-city noises just blocks away.

It is in just such an orderly and spacious place that the seeds of poetry are first sowed. In idleness, and surrounded by natural beauty, children take in, indiscriminately, their first panoramas. They begin to form tentative concepts of how things are — begin to generalize about Mornings and Neighborhoods (before learning, as adults, to turn their backs on such wasteful dawdling because they have schedules, duties, promises to keep).

A courthouse square, with magnolias in bloom and sparrows racing through the morning shade, helps connect us to our past: to timelessness and stability, to other places and other lives.

Grass

It is the grass, the lovely, shining creek grass, that I remember — more than the talk and the cans of Jax beer and the summer afternoon. The three of us — my brother, our doctor friend Lynne Taylor, and I – were driving along familiar farm-to-market roads in central Texas, making a pretence of finding a fishing spot on a relative's ranch but actually just being out in the day, driving slowly in the sun through the pastures. We finally parked at Milton Dyer's abandoned ranch house, walked down the red-dirt slope behind the pens with beers in hand – the afternoon talk still going strong about Dizzy Gillespie and armadillos and Hemingway and Lynne's medical school days in New Orleans – and there was the creek.

We had no real need to fish. We simply stood beside the creek as if we were on a private outdoor stage: the red bluff behind us, leading back to the bright July pastures; the sun low in the west behind the deep-green sycamores across the creek; the grass as smooth and level as a billiard table.

It was Sunday and the hill country was in its last moments of light. . . . What more is there to say? We stood there, we watched small brown frogs sitting on the church-like carpet of grass, we talked, we drank our Jax in the elegance of that quiet creek place until the sun went down and the spell of the summer afternoon was broken and we stood with empty cans in darkness.

Mother

I.

I went to see my mother at the nursing home in Kerrville. I parked beneath the American flag whipping in a breeze above rows of bright sidewalk flowers and entered Hilltop Nursing Home. Old women, seated in the lobby, kept their faces turned toward me as I walked by, but I held no genuine interest for them. If I had been carrying an alligator under my arm or leading an elephant on a chain they would have continued to stare ahead with the same lack of curiosity or concern.

I entered Room 17, Wing D. My mother was sitting in her wheelchair, rocking slightly in the rhythm of her Parkinson's disease and clacking her false teeth. Her white hair was wilder than usual. As

I leaned over to give her a bit of a hug I smelled her Hilltop odor of stale sweat and talcum powder. Her roommate, a stick of a woman hugging her knees, sat in the adjacent bed like a jaundiced praying mantis, staring at us with disapproval. She complained about the Hilltop meals, her stomach, television programs. Bitterness was oozing from her lips like tobacco juice as I wheeled my mother toward the door.

On our way to the home place my mother jerked and clacked and looked out the window. We made our small talk – and I tried to figure out why her words did not quite make sense anymore. It was as though the twisted flow of her thoughts was all that was left of a once-forceful river: one that for 80 years had been broad and smooth but during a storm had finally gone out of its banks. Rock slides had crashed down from overhanging cliffs, tree branches jammed the rapids, boulders filled in the bed. Now, as she talked, the stream that once carried her thoughts disappeared underground, reappeared, detoured suddenly around a hidden ledge, emerged, disappeared into a whirlpool only to reappear once more – a thin trickle among debris.

In the living room I guided her to a chair and leaned her crutch against the wall. I carefully pulled Tiger the cat from beneath the dining table and held him a few moments to calm him. I stroked his back – trying to prepare him for the ordeal ahead. I took him to my mother and placed him carefully in her lap.

Mother clacked and rocked and talked lovingly to Tiger. Then she started to pet him – banging at Tiger's body as if she were dusting a rug. Tiger squatted there for a while – stoically, drawing himself tighter, holding on.

I said, "Mother, Tiger is getting pretty cranky these days. I guess you'd better pet him a little easier."

I went over and tried to show her. I took her misshapen, arthritic hands and guided them slowly over Tiger's head and body, then went to the kitchen to fix her a glass of Coke with ice. When I returned she was banging away again. Tiger growled wildly, dug his claws into my mother's dress, jumped from her lap and ran beneath the couch.

Mother rubbed at the bleeding scratches on her hand and smiled up at me. "Tiger's still frisky as a jailbird," she said.

II.

She was being pushed along in her wheelchair by an aide. The hall was white, her face was white, and her hair, newly brushed, was so white and fine and shiny it looked artificial – a spun-glass wig. She came toward me, unsmiling, strapped in the chair. As I bent down to her I saw that someone, the aide perhaps, had touched lipstick lightly across her mouth in a gesture to encourage life.

It was a pleasant day and I wheeled her outside the nursing home to a porch that faced the sun. I rattled on to her about the hunting cabin, Kerrville traffic, my children, the weather – the useless ritual I went through each time I came to see her. Every now and then she would try to carry on her half of the conversation. She would open her mouth, struggle a bit to shape the sounds, and then speak. Her words were, as before, a Lewis Carroll jabberwocky that lacked meaning – like colored beads selected and strung by a deranged child.

As I stood beside her wheelchair and cut her fingernails I would glance down at her from time to time. She was staring ahead, her face set – and it was as if she were blind. . . . More than that: as if she were both unseeing and unseeable, as if she had disappeared: unreachable within the familiar outline of her body. "My mother" – the woman who had read *Anna Karenina* and Thomas Mann – was gone; yet, incomprehensibly, she was still there: a white-haired, white-faced, lip-daubed figure in a wheelchair: a stranger staring into a landscape she could not describe, thinking thoughts that had no language.

III.

In March I had given her a quiz. "Who am I?" I asked and gave her four names to choose from. To her, I was none of them. She was comfortable with the familiar sight and sound of me in the room, but I had become nameless. She was beyond motherhood, beyond the bonds of human attachment; she could no longer conceptualize me as her son. She lay on her incontinence pad, beginning to lapse into a coma — the final mystery close at hand.

In May, at age 84, she died. At the funeral home where she lay in state family friends came by to sign the register and pay their respects. They solemnly shook my hand and said It Was For The Best, It Was a Blessing. I suppose they were right — about death being the proper thing for an old woman whose brain had slowly been degenerating. I had no better stance to take toward death as I stared at my mother in the casket. She was now a cadaver made ready for burial: artfully powdered and painted, bled and disemboweled and filled with formaldehyde. Her arthritic, knobby hands lay crossed on the funeral dress, her white hair gleamed, sprays of flowers were arranged about her in a semicircle. My mother, Maggie Mae, was ready to be placed in the ground.

Her funeral was at noon in Harper. As cars pulled off the single main street and parked on the grass of the small wood-framed church, a mockingbird sat on top of the church roof and poured out its dazzling sounds. The service began: the pianist played "In the Garden;" the soloist sang "Ramona" — the song my mother had sung to my brother and me when we were children; the minister recalled for us my mother's life and her virtues; another hymn was sung. The service ended.

We drove to the cemetery across the road, the minister said final words under the canopy, and my mother in her closed casket was poised over a hole that had been dug that morning next to my father's grave. We drove back to the hall behind the church and had sandwiches and iced tea.

. . . And I think about it: how I had stood looking at my mother's profile in that darkened mortuary room in Kerrville — faint organ music constantly playing — and half-expected to see her move; how people had moved decorously across the carpet toward the casket and looked at her and said "She's better off;" "She looks so peaceful;" "They certainly did a good job" — as if she were merely asleep or in suspended animation . . . how we, the living, carry our own life within us like an invisible candle flame that will never flicker out . . . how we live intimately with both life and death and do not understand either one.

IV.

I stood by the mound of my mother's grave in the small Harper cemetery. She was down there in the dirt. A week after the funeral she was still stretched out under ground.

"Hi, ma," I said.

I was properly erect on a June day, standing the way living humans do, and she was prone in her casket, getting ready for the bugs and worms.

Two feet away was my father's tombstone (1895-1981) and, beneath it, his skeleton. I looked around: I was in a community of gravestones with their carefully chiseled names and dates. And underneath, in their assigned portions of earth, were the Harper ranch people who once drank coffee on their screened-in porches and discussed the price of Johnson grass hay. My mother had just joined them — a newcomer in her dusty-rose burial dress and shiny green casket.

As I left her mound and walked past the grave sites toward my car, my legs and arms moved with automatic ease. They knew what they were doing. I was unquestionably alive — putting my familiar body into the familiar car so I could continue the familiar, normal business I had briefly interrupted by stopping at the cemetery. I was not dead; I was not under dirt. . . . As I drove down the lane toward the highway — past the Harper school building and the school buses dramatically yellow under the noon sun, past the shining oak trees, past the radiant roadside summer flowers — I could not fit the two together: being dead, being alive. There was no way that I could bridge the gap between the silent cemetery world and the sun-bathed morning. Those names carved in marble — Hopf . . . Bierschwale . . . Tatsch . . . Ahrens . . . Sageser — and the bone-people beneath: they simply did not belong to shining oaks and yellow school buses.

Cars kept passing me as I drove on slowly toward Fredericksburg, and I watched them disappear down the highway: all those people on the move, in a hurry, full of goals and purposes, with their children gazing blankly at me through their rear windows — children I would see once and never again.

What could I do about them: fire and ice, life and death, *I-thou:* always separate, never merging, never knowing the other?. . . . I looked at the pasture trees: I would continue to spend my life next to them and not know how it was to be a tree. Or a dog. Or any other living thing. . . . We lived in compartments – fractional, isolated. I would only know me and my life, and then, one day, I would be in a box in the ground and my son and daughter would stand there, six feet above me, in the sun, and say, "Hi, Dad" and I would not hear.

AT THE HUNTING CABIN

I.

I reclaimed, last summer, a personal territory, a place to call home for a while. I lived at the family hunting cabin among the trees – wanting nothing more than what I saw before me each day: pastures, sunlight, hillsides. My cup was empty when I went there in June, but by August it was full.

Nothing flashy or dramatic happened: no meteors dropped from the sky. I simply lived as though I were seated on the front row of the First Day, watching Creation. The countryside was my Book of Knowledge.

When I woke on those overcast mornings in June I had to adjust to my new reality. It was 7:30 by my watch, time to be headed glumly to work on a conveyor belt of cars, but I was no longer on Loop 410 in San Antonio. I was lying on my sleeping bag with a delicate south breeze bathing me through the cabin doorway.

I had no plans to follow, no obligations to meet. The day was simply what unfolded: read in the shade of one of the oaks, look at gourds growing so pleasingly striped and round and green among their vines, watch a buzzard soar, a cow chew. I could cut cedar if I wanted to be productive; I could drink cool water from the windmill if I wanted to feel contentedly alive. It didn't matter what I did or did not do. Human *doing* was not on the hunting cabin agenda. It was the natural world that was on display there, and I was ready to be its dedicated observer.

After luxuriating in the breeze that touched my face and arms as tenderly as a geisha, I made coffee on the Coleman stove and sat in the doorway in my bare feet. I drank coffee while a wasp investigated the window screen and lone doves sat on the bare limbs of a dying oak – as if resting in the somber branches of time. I looked at lichen-covered rocks by the cabin gate, then at the liver spots on my hands: unnoticeably, we had been aging together like country cousins.

II.

Each morning the cows came into the water lot. They convened slowly from the various draws and hillsides in a carnival of calls and bellows. They announced themselves without preamble or niceties, their sounds abrupt, strangled, imperious: *"I'm* here; where are *you*?"

One of the Herefords stood beneath the huge water lot oak, the peaceful captain of her ship – while her calf, a big one, sucked and jabbed with abandon. After finishing his assault the calf drew his head back and stayed there at the mother's side, nose to hip, facing west. The mother remained motionless, facing east. As other cows drifted past, mother and calf stood side by side, solidly content, as if in harmony not only with each other but with the earth itself – indeed, with all heavenly bodies.

One morning three cows came down the arroyo, around the corral fence, into the water lot – slowly making their way through the tree shadows like late arrivals at a camp meeting. They stopped at the water lot gate, looking distracted; then they raised their chins and bellowed fiercely toward the pastures. They were bereft, you see. Their owner, Talbot Garven, had come to the cabin the day before, sprayed the herd for summer flies, and afterwards had loaded eight calves in his trailer and taken them off to market. The three cows at the gate were several of the disconsolate mothers, wandering separately – grazing their way through the early morning, stopping, staring, calling out.

The cows at the gate were studies in depression, unwilling, even, to lean down and pull off the inviting top of a nearby purple thistle while they waited. One of them finally shook free of the mood and tentatively began to graze. But in a moment she raised her head – not caring that she looked wild-eyed and dumb as a stump, with slobber and runners of mesquite grass dangling from her mouth.

In the afternoon I sometimes opened the gate of the cabin lot and let a few of the cows in to graze on the thicker, greener grass. It was like a summer lawn party, with the invited guests on their best behavior: ten Hereford cows, all in red, and Bull, their dehorned, plodding companion. Several cowbirds always flew down to join them, and like small French serving maids moved briskly among the hoofs and heads, sometimes even busybodying their way beneath a cow to get to the grass on the other side — as if hurrying beneath an arch on a familiar Paris street. The cows, unconcerned, grazed along in their green commons — past the outhouse, the rolls of wire, the discarded harrow, their tails making a casual swinging ballet in the bright four o'clock air.

III.

At noon I sat in the sun, my body hot with life, and I wanted to sing the summer as a song, the way the locusts did. I wanted to turn the day into words, but everything important was being said: the sun and the earth were saying it all.

Beneath the big oak in the water lot cows licked salt blocks; others stood at the water trough, drinking in long smooth throatfuls. Finally, through with licking and drinking, they stood in the shade, chewing, looking out.

The cows and I were witnesses of that midday hour: We saw yellow butterflies sit delicately on moist manure piles — like trim Chinese sailboats becalmed on muddy flats of the Yangtze.

We saw the weathered boards of the cabin lot gate gradually darken as the trees drew shadows over them as smoothly as a curtain.

We saw the sun gloss the smooth poles of the corral fence, making them shine like the calluses of a worker's hand.

We saw great sagacious trees hold private conferences along the hillsides — the sun scattering highlights on their loose green robes that flowed and lifted in the noon wind.

We were witnesses and I was awed, even if the cows were not. I waited patiently for meaning to emerge from the stillness of the rusted oil drum and the cedar posts in the fence. I expected no less of them than Buddha might under his banyan tree.

I was trying to slip past nature's guard, you see. I looked and listened — I was attentive to the processes going on in every square and cubic yard of my cabin world — and I wanted to get a glimpse of what was happening, so relentlessly. . . . How did it work, this incredible earth? I wanted to solve the mystery of the commonplace — come to grips, at last, with the tantalizing oak tree shade that lay like an eternal hieroglyphic against the rocks of the hunting cabin wall.

I looked at the sky through the burning glass of midday and thought of Greece: didn't it have the same heat, the same locusts singing from hidden places, the same bare, white rocks and caliche roads? When I walked up the arroyo in the late afternoons to my favorite shelf of rocks and sat beneath the gigantic oak among its bare, Medusa-like roots, didn't I ask it my unanswerable questions and didn't it, like a hill country temple of Delphi, offer its silent replies?

IV.

The Spanish goats started coming by in the early afternoon, browsing their way north through fences, across pastures. They belonged, apparently, to a neighbor near the interstate.

On the first day they appeared the cows were taking their ease in the front clearing. I was reading in the cabin doorway. The goats, nine of them, simply materialized in the patch of purple thistles by the molasses wheel.

I watched them: nicely brown and black and white, a pleasant and unexpected plus on the landscape. But since Spanish goats are generally a stand-offish and independent breed — gypsies of the pastures — I assumed they would just give me a look of casual disdain and move on about their business. In a moment, however, the smallest goat, a black kid with a white face and lank belly, came bleating up the fenceline. He headed toward my outstretched hand and began sucking away at my thumb. He had definitely been a bottle goat but I had no milk for him, not much of anything store-bought except carrots. I got one from a cellophane sack in my cooler and offered it to him, but he was not interested. He wanted milk.

It was then that I met the black billy. He had started up the fenceline too: his smell, like a pennant, announced him. The kid dodged to one side as the billy, with his steady black eyes and dramatic goat smell, walked forthrightly up to the cabin. He regarded, circumspectly, the carrot in my hand; then, satisfied it posed no threat and might, indeed, be as interesting to eat as to look at, he leaned forward, his upper lip curved outward like a black pseudopod of flesh. I cut the carrot in two and offered him a half. He took it: bit it and chewed it solidly, and waited. I offered him the second half. He crunched it too.

That was that – like the first hit of heroin. The black billy was hooked.

The next day, at noon, when the goats were back again among the thistles, the billy licked at the molasses wheel a while – then bleated. I was out of sight inside the cabin. He licked and looked toward the cabin and bleated some more.

I was bending over the cooler when I smelled him. He loomed in the doorway like an apparition from child's nursery book and watched as I cut slices longways from a carrot. We went outside. I fed him two carrots, slice by slice, and then shooed him back to the water lot and closed the gate. He reared up, put his forelegs on the wire fence – leaning toward me as far he could. He stretched forward, the tip of his tongue reaching out in shameless little quivering, searching thrusts. He was a goat dangerously close to losing his dignity. I had one carrot left in the cooler and I gave it to him. He chewed it while still leaning awkwardly over the fence.

When the rest of the goats began to ease around the corral and on through the trees, the billy lingered. He stayed there with the cows during the midday heat – nibbling thistles. The Herefords moved out to graze in the late afternoon, but the billy remained. At five o'clock, when I began to make coffee, I could still hear his tentative bleat: the half-hearted cry of a cold-carrot junkie, desolate and alone on his hillside.

V.

At nine o'clock the ranchland was still covered by low morning clouds, but the birds were singing away as if already drunk on sunshine. Small black squirrels strayed from their home in a walnut tree and climbed within the squares of the cabin lot's wire fence: squirrel children taking turns on back-yard monkey bars. On a nearby knoll a deer and fawn stood like light-brown statuary placed overnight among the trees.

In the clearing a cow licked her calf – newly born, with some of the new still on it. The cow licked along one side of the head, then along the other; along the top of the head and then under an ear. When the cow finished with her own calf she turned and started in democratically on a nearby yearling, who stood luxuriating in the rhythm, the touch as the licking went on and on.

The wind rose, the windmill turned, the leaves of the Spanish oaks – pale green and papery and delicate – jiggled and rustled for a while. The wind was like a surf among the trees.

A long spear of pasture grass waved and bent in the air – a yellow-tipped metronome – and the cabin lot could have been a Dakota prairie in the 1800's or an African savannah with small, pre-human hunters stalking game with pointed sticks. The land seemed as young as the morning, coeval with the rabbits and black squirrels and the birds.

I knew I would accomplish very little that day. Yet, I wondered, what was the proper measurement to use in gauging work? If it took nature 500 million years to lay down a new stratum of rocks in the Grand Canyon, what should nature – or I – be doing on a Tuesday?

I looked at a weed in the water lot. It was an ordinary weed but nicely shaped, as weeds go, with several yellow flowers. I could chop it down and feel productive or let it remain there in its bare space. . . . I looked at it closely: it had a nice completeness, one might even say style – certainly a context among the grasses and animals.

I did not touch the weed. It belonged in its spot of ground as much as I belonged in the cabin doorway.

VI.

As I read on the front porch slab a mouse came out from a crack in the wall. He was a wee beastie, as Robert Burns would have said – about half the size of my thumb – and his long trailing tail looked like the queue of hair that young men wear at the back of the neck.

At first I thought he was just looking for adventure: a mild, baby-mouse kind of exploration along the side of the shaded tin wall. But after watching him a while I understood his mission. He had eaten my Decon-poisoned grain and was on his death march.

The field mice had been invading the cabin at night and I had finally decided enough was enough: I had put Decon paper trays under a table and waited. The mice had been making themselves far too comfortable in my sack of groceries – one had even become a fan of chocolate chip cookies. (I would lie on my sleeping bag in the dark and listen as he rattled around in the sack – gorging himself. When I would shine my flashlight across the cabin he would appear for a moment at the top of the sack – hanging there with a kind of bemused, tolerant expression – then go right back down to his midnight feast.)

I put my book aside and bent down to look closer at my porch visitor. The little guy would move and stop, move and stop, not knowing what the devil was giving him such a bellyache (. . . "Is this the normal way to grow up and be a field mouse, Ma, or *what*?"). Not mouse-like yet in shape – not sleek, not ready for climbing and cavorting – he was still just a hump-and-tail, a kind of tadpole-mouse edging himself along the cabin wall.

When I got down on my knees I could see that his eyes were closed – either he had shut them against the death agony or had not yet opened them after birth: he had never seen the world he was going to die in. He had begun to tremble and twitch, like a little old man with bad dreams nodding in a chair. As the sow bugs – gray, relentless 18-wheelers on their personal interstate across the concrete floor – passed over his tail, the mouse would flinch, blindly, then face in another direction and sink back into his private quivers. Every now and then he would get an itch and would scratch himself with his perfectly shaped little field mouse feet – lightning fast, it seemed to me, for so comatose a creature.

After a while I got up and went about my business – patching a few holes in the cabin roof, straightening up a wire fence. I checked on Mouse once to see how he was doing, but he was gone – had bumped his way back into a crack in the wall, I assumed, and was somewhere beneath the cabin.

As noon I went to the windmill in the cabin lot, got a long drink of well water, put out salt blocks for the cows. When I came back the mouse was lying on his side in the sun. A very still little body on the hot porch slab, he had been made dead by me. I had sowed my seeds and reaped this small harvest. He had taken the poison-dusted grain that his mother – good provider that she was – had brought back to her children and – good son that he was – had dutifully eaten it.

Somehow, stretched out on his side, he looked larger — more mouse-like in shape — and had an impressive kind of mouse dignity: front feet formally together, touching one another; ears flattened; fur neat; tail curved down in a proper symmetry. And of course the eyes were still closed: Mouse's brief span of days had been completed in darkness, and he was unaware that he cast no shadow in the blazing summer noon.

VII.

I grappled with my problem of the moment: how to interpret a cow's stare. That is, how could I get down on paper the essence of a Hereford's stare as she lifted her head at mid-morning, turned slightly, and regarded me from within her solid cow center? How to convey that deep cow-superiority and cow-indifference, that all-seeing and unseeing gaze-and-drool? The Hereford's look — apparently so guileless, so impersonal, so much like a blank pane of glass in an empty house — seemed, suddenly, to contain an awesome, cryptic message coming to me across the chasm of evolution.

The Hereford, drinking at the trough, dripped and stared, and I waited for a breakthrough. I demanded communication: I wanted to see past her pink comedian's nose into a hidden First Cause.

I moved towards her. I walked slowly, stopped a moment, talked to her quietly, came closer, stopped again. I was within six feet of those childish, towheaded swirls and curls above the blunt plain between her eyes. She continued to stare at me from beneath the awnings of her long, white, girlish lashes and then — convinced that I was not hiding feed for her behind my back — moved off heavily to the salt blocks, imperturbable as Nero.

VIII.

We were together in the heat, the cows and I. We were under our respective trees, looking out together toward the bright August afternoon. A buzzard drifted overhead, a deer crossed the road at the top of the hill, flies angled about, bees settled and lifted from the water troughs. In the rock tank frogs sucked their throats like small contented boys rubbing balloons.

It was hot under my oak tree, but I did not abandon it. I breathed, I existed, I rested like the other creatures of the earth. There was no hierarchy among us. We shared the given moment.

In the late afternoon, as if welling out of the heat and the waning sunlight, a dove began calling. It was like an escaped breath from the land itself, an unobtrusive, suggestive voice saying "Do you remember? Do you still care?" Like a tide coming in to shore, the dove call flowed so evenly through the trees it was as if I had thought the call rather than heard it.

I followed the call, somewhat. I walked toward it down the dry creek bed west of the pens, then I looked back to the clearing of the cabin and the remains of my father's old feed store truck there beside it. As I stood beneath the walnut trees, on the layer of smooth rock where we had once had family picnics, I thought: How do you repay a pasture, a clump of trees, the sun? What can you give them for making your life worthwhile?

The dove called, again and again, taking me deeper into the mood of the silent rocks and lengthening shadows. As night fell, it seemed to me that the only way to give a proper thanks — to the earth — was to keep on paying it close attention. To keep looking hard and saying: Yes, I see.

IX.

"Chi-ink. Chi-ink."

Gregorio was digging post holes in a layer of rock.

"Chi-ink. Chi-ink."

I saw him on the fenceline morning and afternoon — blue cap, red shirt in the sunshine-and-shade. He maintained a smooth, swaying little rhythm, almost a dance step: feet in place, swinging a bit from side to side as he sent the crowbar twisting down.

Each morning he got up at six, made breakfast in his tent, and was there at the fenceline as I rolled over in my sleeping bag outside the hunting cabin door.

"Chi-ink. Chi-ink. Chi-ink. Chi-ink."

At noon he would stop and come back slowly toward the pens, walking past his row of cedar posts gleaming ivory-and-brown in the sunlight. He would eat and rest in the tent, then walk back to his crowbar and gloves.

"Chi-ink. Chi-ink."

The afternoon shadows deepened, the sun lost its glare, doves scouted above the walnut-tree arroyo, and Gregorio worked at the fenceline, cutting cedar trees, stripping and shaping posts – aligning, tamping them until they were solidly in place: as if made of rock themselves.

He would stop working just before dark and come to the water lot to fill his bucket at the windmill pipe. Sometimes he pushed back his cap – at 53, his hair remarkably full and black – and smoked a Bugler cigarette and we talked about the cows, or cedar ticks, or the spread of purple thistles in the pastures. He spoke deliberate and carefully enunciated Spanish.

Once, when we were talking about sports, I mentioned that my son had become a cross-country runner at his high school. Gregorio talked at length – almost didactically – about fitness programs for athletes and the Olympic trials he had watched on Mexican television. He said his sons had also been runners in school. When our conversation died away and he finally lifted his bucket and headed toward his tent, the bullbats had begun to sweep over the water lot and katydids were flashing their messages in the darkening trees.

. . . Gregorio: I think of him and the other lives he might have led. I see him seated behind a desk, an investment broker giving measured, articulate advice to a client. I see him in the dugout of a semi-professional baseball team, his silk manager's shirt fitting tightly over his thick chest and muscular arms.

As I drive down a city street or have coffee in a cafe, I hear his crowbar biting into rock; I see oak tree shadows sliding across his blue cap and sweat-circled red shirt. A cardinal flashes through the cedars, a buzzard coasts in the mid-summer sky, and Gregorio makes another envelope of dollars he can send to his sister in Piedras Negras.

"Chi-ink. Chi-ink. Chi-ink. Chi-ink."

Alone in the pasture, from daylight till dark, Gregorio digs through hill country rock. I listen, and he is like a myth that keeps growing in my mind.

JOURNALS

Writing is directness. It is the body opened, the bones bared.

Peel off the hide of a person doing his job *competently* and you will find a person secretly sick at heart. For it has ever been so: that to do a job merely well is a kind of death. It is only when a fresh wind comes upon the man's blind side and catches his competency off guard – suddenly touches the skin hidden beneath the accreted shell of skills, poses, wearisome routines – that he stands the slightest chance of rejoining life: of knowing once again, as he did in childhood and youth, the miracle of doing a thing because he likes it, of doing something out of joy.

I believe that hummingbirds and cedar posts are better for the soul than all the hymnals in the Methodist church.

If Tolstoy had been a Texan and if he had written a *War and Peace* about his native land in, say, 1958 but then did not write a single line thereafter, would he, today, still be talked about in reverential tones – would he still be considered the greatest writer in Texas even though he had not kept on *producing?* Is continuous, or current, production the yardstick used for literary assessment? Would critics carp at him for Failing to Live Up to His Potential?

Art occurs when the passion of living becomes too much to bear. If there is no passion, there is no art.

The art of writing is somehow locating those concerns which appear in the mind as fruit ready for picking. Such writing is merely a curious kind of play – a wandering among the fertile groves of memory, pulling a peach here, a plum there, reaping a harvest of words out of the rich, shadowy orchards of the past.

The point is this: Don't try to create the fruit; don't struggle to grow it by craft. The apple of an old sensation appears out of its own mystery – not by command.

Austin days, 1954: Full of bright slanting sunlight and tree-cast, sidewalk shadows and the glaring sides of brick buildings absorbed rather than seen by the walker's eye. Long stretches of green lawn and blue-white sky and lazy midday heat. Hackberry trees hovering over silent, unused back yards. . . . And in one such yard, on the back porch of a boarding house, a black cook squinted up at a loud airplane going by: standing there, a huge, glum woman in an apron, shaking over and over the last drop of water from a dishpan before turning to go back inside the kitchen – into the slavery of her pots and pans.

. . . You walk the sunlit streets and come to a filling station rest room where, after taking a leak and washing and then waiting for the paper towel to pop out to dry your hands, you make a quick evaluation of how the day – that is, your life – is going and you find that it is not going well at all. Outside, you pass old men on the corner, fumbling with something inside the labyrinthine compartments of their heavy clothing. You stand there, looking around, and find yourself facing, across the street, the doorway of a narrow, darkened mid-street bar. Out-of-work men wearing last week's shirt are aiming toward it as they cross unsteadily against the traffic light. You walk away, walk on – knowing that the purpose you so desperately seek for yourself is not to be found in the neon-lit gloom of The Friendly Tavern.

Do you know what a "sense of self" is?

Well, drive sometime along a pasture road in the hill country of central Texas and come across a Hereford bull planted solidly on his four feet in front of you in the roadway. Study that bull as he stands there. Notice he looks at you from deep within his bull-ness and feels – with absoluteness, with finality – that he has no need to move out of your way. There is no bell ringing inside him that signals, even dimly, the possibility of danger. His DNA is light years away from registering you as a threat.

If he could speak — as he looks at you without curiosity in your wraparound carton of metal — he would say, unblinkingly: *"Hunhh? Wachu want?"*

That, friends, is sense of self.

Sometimes a person puts down phrases one after another about a certain thing he has in mind. He then considers his work a moment, perhaps crosses out the first five lines, substitutes two others, then after reflection adds a single word more. He is engaged in what is called the conscious act of writing.

There are other times, however, when the putting down of words is no more like *creating* than holding on to the neck of a runaway stallion is like horseback riding. The person grabs blank checks, old grocery bags, whatever is available to scribble on, and with them in hand dips into the stream of emotions that is suddenly flowing by. And as the words hit the paper there is no crossing out, no pausing to consider or reflect — for the moment is really not suitable for writing, for crafting words. The person simply realizes the dam inside him has truly burst and all he wants to do is try to find enough empty buckets and coffee cans to catch the swirling waters of memory before they flow out of sight forever.

There is nothing more energizing than unquestioning belief: belief in (school) (neighborhood) (religion) (nation). Think of all the Mormons, Baptists, Muslims, Communists, Central High Mustangs, Americans, Brentwood Park Development Association members, and the like: they get things done. They work — work hard — because they have a context. They *believe.* They are protective and suspicious of. They are *against.* They are convinced of the rightness of their position.

Those of us who cannot always be convinced that our (neighborhood) (religion) (nation) (race) is the (best) or the (answer) are the loners and drifters — paralyzed in the marketplace of ideas. We are unable to believe 100 percent. Our passions are undercut by our lack of conviction that being born, say, into a Catholic family in Ireland or a Shiite family in Iran has anything to do with Ultimate Truth.

(Belief) and (Faith): they vitalize. They bring people clean streets, a readiness to kill and a willingness to die, a sense of pride in (Ours), contempt or hatred for (Them and Theirs).

I remember when everything was real in my back yard. Tree shadows were still as good as they were when I first saw them as a boy. The moving November air still carried, through the Sunday quiet, the sound of neighbors settling spoons on dinner tables and car doors slamming on nearby streets. My back yard encouraged sunlight to remain on a white duck's back; it sustained the deep immobility of a shovel lying on the ground.

I sigh, and think of other times. I remember teaching on the Gulf Coast – and a girl, of course. She was Mary, and we were together then. I think of the grassy yards of small Baptist churches near the bay, and late Sunday afternoons with the smell of salt water and coastal grasses in the air. We were companions that year, Mary and I. We were in our twenties and thinking about marriage but not quite sure that marrying was what we wanted. So we went a lot of places together – to her relatives to sit in their farmhouse living rooms and on their front porches and listen to their idle talk; to Corpus to drive along the sea wall and feel the blackness of the coastal night; to small cafes along the highway to hear Lefty Frizzell tunes on the jukebox and drink slow cups of coffee and, on the way home, to stop and hold one another in the deep South Texas darkness.

I have never quite accepted the stillness, the unresponsiveness of trees, posts, rocks, grass. I am still waiting for them to speak.

I am so conscious of them that I cannot believe they will always remain silent. I keep being watchful, alert, listening and waiting, half-hoping to overhear the sound of their quiet afternoon voices.

I have tried to set down, memory by memory, the meaningfulness of who-I-am: to record the snail-slick of my passing, the briefly charged intimacies of my days, the residues of my private fires.

The only truth I am interested in is the truth which I can reveal to myself and which I can translate into my own words.

Imagine: a happily married couple. Just that: two humans, delighted with each other, pleased to be in each other's presence.

It is a condition too amazing to believe.

Not doomed lovers of twenty-two but a man and a woman in their forties or fifties — still glad to touch each other's flesh, still glad to share one another's daily lives and routines, still needing each other, still functioning best when the other is around. . . .

Writing? It is pleasurable, it is private, it is one's kingdom by the sea.

It seems that everyone who fools with words, at one time or another, climbed to the top of a hill overlooking his home town and gazed down on the scene below. Standing there, he becomes aware not only of the love he once felt for his boyhood surroundings but also of the loneliness and perspective that the hill affords.

I need fields and fences, corner posts and cedar stakes: white-faced rocks lying on bare, rain-wet winter ground. I need oaks standing about in groves, somber as committees, and barns at a distance on a rise of ground. I need hot tin roofs shining at a turn in a country road, with streaks of rust making a casual rural elegance.

Some day I would like to write a small book as sweetly done, as satisfying as *Cannery Row.* I would like to write it and have someone, on a pleasant afternoon, say to himself, "I think I'll put Bode's book in my pocket and take a walk. . . . Maybe I'll stop at a bench somewhere or a deserted porch and read a while in this book that I have come to love."

I wish that would happen because I have done that very thing with Steinbeck — have taken his book and started out somewhere with it in my back pocket and have stopped for a while and read in it for the sixth or seventh time. It is a friend of mine, and as we have sat there — under a tree, on a rock wall — I have read and then looked up at the nearby buildings, or at people walking by, and have felt that I was in a very fine place. The book had made me feel that way: it was a link connecting Steinbeck and me and the world. We had a pleasant kinship, one that made me feel at ease with myself, that made me smile.

In the riverbottom at Vanderpool: Birds moving slowly about among the huge cedars, not singing, flapping their wings heavily like women popping laundry. Cardinals, woodpeckers, robins down south for the winter. There is no wind. Sunlight filters through the top branches, there is the clean smell of the cedars, the Sabinal River is running nearby. There is a deep quiet, a sense of isolation and grandeur.

Rural, smalltown men aren't given to exaggeration. Had one of them seen the Great Flood of the Bible, he would have reported to his buddy at the coffee shop the next morning: ". . . Purty good shower over to Noah's place yesterday e'nen."

I remember my grandparents driving to town through the pastures in the two o'clock heat, the white caliche dust rising in an almost heraldic plume behind their green Buick. Gram sat in back, her hair pulled tight from the part in the middle; traces of spilled face powder were on her silk dress. She was neat and compact, and her rimless glasses caught the light whenever she turned to look out to the pastures. Grandpa — profiled and Stetsoned against the window — sat solidly behind the wheel, his boots black and shining on the floorboard.

It was Saturday afternoon and they were bringing me back home after a week's visit to the ranch. The Buick moved like a ship through the sunshine and the heat, past the raised heads of the tree-shaded sheep, over the rattling cattle guards, down the road toward the ridge of Kerrville's blue hills.

I want to talk about the ranch, over and over. I want to tell about the smooth board that Gram pulled out from the counter by the sink and sprinkled with flour: the biscuit board. She would stand there making biscuits while she talked to mother — the kitchen yellow-glared from the bulb hanging down and warm from the huge wood stove. She would be there in her apron near the door that led onto the small back porch, the square pane of glass in the back door black with night — with ranch night. . . . Gram, the fleshy part of her arms jiggling a little at the elbow as she rolled the dough and flour.

The ranch kitchen: cabinets painted yellow enamel, the flour bin that yawned outward from the floor, the flyswatter hanging from its nail, the smooth, worn linoleum.

I want to go around, touching with words the remembered objects at the ranch. It would not be writing, creating; it would merely be, if nothing else, bidding a proper farewell to things of the past.

Is that what we do: live, experience, and then, if we are driven to do so, give things a proper burial? Is that all we *can* do? Live, use, bury?

A person writes from the bed of hot coals within him — some fierce and blindly furious quarrel with the world. If he uses the heat of his quarrel to mend iron words into place, they retain their force long after the original fires are cooled. A work without such moral passion is doomed. It loses its glow and becomes just another meaningless lump on the discard pile of literature.

Any one day is perfect and all the things that are in it. Any one moment is infinity, eternity, happiness — what you will.

Most of us cannot experience infinity, or happiness, because we are too much a part of the flow of time. We do not try to stop it.

Mysticism and poetry are just that: stopping time, getting outside the seductive blur of its passing.

Perfection, you see, is this instant. Bliss was possible a moment ago.

Life never becomes greater than what it is right now, but we always keep looking for *more.* We are carried along on the conveyor belt of time — dreamily, day by day, year after year, until finally the conveyor belt stops for us and, shocked, we yell out: Wait! Let me go back — I'll get it right this time. I'll know what to do on the second go-around. . . .

But it's too late and so we die — crossing our fingers, hoping for a paradise like the one we had just ignored.

Two scenes: It is a summer noon and I am walking as an eight-year-old on a dusty, unpaved street in Kerrville. I pass a small, yellow-frame house. I look at the yard and then at the screen of the bedroom window. A portable radio sits just inside the window screen, and through the dimness of the screen I can make out, in the semi-darkness, an ironing board and the arm of a woman as she irons.

As an eight-year-old I do not know that I am living during a time called the Depression. I think that it is simply our neighborhood I am in and Mrs. Latta is ironing Mr. Latta's khakis and the clothes of her six children. I watch the dusty screen and the arm moving slowly behind it. I can hear the sound of "Ma Perkins" coming from the radio. I can smell the cabbage from the Latta kitchen. I see the tricyles upended on the front porch, the chinaberry tree with its cluster of pale yellow berries. I feel the noontime heat and hear the locusts surging and shrilling in the trees along the other yards of Gilmer Street as I walk on up the hill toward home.

. . . I am 17. Supper is over, the sun is down, and my mother has resumed her watering in the front yard. She is almost lost among the vines and lilac bushes.

For long moments she stands there, paying no attention to the garden hose in her hand or to the water that is making silver foam in the dirt. She is looking past the flower bed to the vacant lot across the street. She is deep within herself.

I come across the yard, ready to tell her that I am leaving — that I need to go back to school to practice for the senior play. She turns, looks at me, smiles. (It will be years later that I will read the line in James Baldwin — "You take your worth from other people's eyes" — and realize that I took my worth from the eyes of my mother.)

Two scenes among a thousand or more; and as I sit here, years later — writing at a table on a windy, winter's day — they are still as real to me, as immediate, as current and significant, as the air I breathe, as the sound of the typewriter keys hitting paper.

Through a Glass Darkly

I got out of my car in front of the Riviera Restaurant late one afternoon and started to go inside. I looked idly at a nearby street sign, the dust in the parking lot – and suddenly the whole routine business of daily living fell apart. It was as if the universe had quietly stalled there on Doniphan Drive and I was looking at life with X-ray eyes.

It's hard to explain this kind of moment: the unexpected refocusing that allows a person a quick glance at what he ordinarily does not – cannot – see. It's not that the street sign or the nearby telephone pole or the bits of gravel in the parking lot had come alive. But it's *almost* as if they had – as if reality had casually allowed its mask to slip, had allowed me to see objects and surfaces in a different, more timeless light. It was like looking up – expecting to see a friend's familiar face, body, clothing – and staring instead at the skeleton, tissues, cells inside.

As I stood at the Riviera entrance, I thought of a similar moment the previous summer. It was midday in central Texas and I was driving through a sunlit pasture. Suddenly the trees on each side of the road were so real, so electrically vital – so charged with life – that they became *unreal:* they lost their everyday, commonplace masks. They were no longer just cedars and oaks – familiar objects with familiar names – but tree-shaped mysteries, new and awesome.

What is it that we actually see – and what is it that we assume we see? We look constantly, and each day our human eyes report to us a human world: a reality separate, however, from that seen by grasshoppers or dogs. What we agree to look upon casts a deceptive glare.

Perhaps we remain complacent – or sane – only because we see imperfectly. Perhaps the intensity of a single creation clearly perceived – an atom, a moment of time, a tree – would be too much to bear. We could not stand to gaze fully upon it.

Mysteries

On a summer night, some years ago, I was standing on the front porch of the family home in Kerrville. It was two o'clock in the morning.

Inside the house everyone was asleep. The arms of my children were flung out across their beds against the heat; oscillating fans were whirring. Outside, the trees in the yard were quiet accomplices of the night.

At first, as I stood on the porch in the dark, we had never seemed closer – the night and I. It did not seem galaxies deep. It was as personal as humans have always wanted it to be. . . . The night? Why, wasn't it just those familiar bushes along the fence, neighbors asleep nearby, unseen streets of the town? Wasn't it just ordinary shapes recast for a while in darkness and shadow?

It was not. It was my nemesis: the taunting, insect-pulsing riddle of my adult life. Yes, I was standing there on the porch, but the night did not know me and I would never know it. The night: as close as the trees and the moonlight but, like the universe, unfathomable, unapproachable.

As I listened to my children turning restlessly in their sleep, there was no difference, it seemed to me, between the mystery of their dreams and the mystery of the all-pervading dark. Both lay beyond human knowledge – had their origins beyond the stars.

Imagination

Imagination isn't what people think it is. It is not imagining what doesn't exist; it is imagining, intensely, what *is.*

Let's say I am walking down Avenida Juarez and I glance through a window. I see a Mexican girl behind a counter selling glass figurines to a customer. As I pass on by I happen to look down at her shoes – and imagine her bare feet within them.

I am not thinking of something futuristic, escapist, exotic. Out of an interest in the girl behind the window I am simply wondering about her and the life she leads as she bends forward over the display case, her hair across her shoulders, her feet bare in her narrow shoes.

Imagination is part curiosity and part concern, empathy, feeling: a desire to enter – to penetrate – the unknown. It is not an unearthly flight from the real; rather, it is an intense desire to extend the senses: to see what the limited eye cannot, to flesh out the skeleton of merely personal experience, to broaden the human range. Imagination, at bottom, is the unconscious effort to be God-like: to know all that there is and be part of it.

Intimates

On a bus trip through New Mexico a woman in front of me, almost asleep, had leaned across two seats. The top of her head showed above the brown plastic seat cover and her bare foot stuck partly into the aisle. It was the nicely shaped foot of a woman in her thirties, with ordinary blue veins beneath ordinary white skin. As she nodded in and out of sleep, she lazily flexed her foot, spread her toes.

I believe I was the only passenger on the bus who was aware of the woman's toes, that milky skin. We became intimates, her foot and I, from Sante Fe to Raton.

I kept thinking: I want just such a woman's bare foot in my daily life. I want a woman and I want her foot lightly sheathed in a brown sandal as she walks along beside me. I want to see her foot there on the living room rug at the end of the day. I want to see it protruding from beneath the sheet of our bed at night.

But at Raton the woman on the bus took her foot from me and disappeared into the waiting room. She did not come back. I was left alone with my book and the stale smell of cigarettes all the way to Walzenburg.

Moments

Significant moments in one's life can happen unexpectedly, strangely.

I remember standing one night at a hotel window in San Antonio when I was a boy. The light was off in the room and as I stood next to the window, with the summer breeze drifting in, a train whistle blew somewhere on the horizon.

Nothing I had ever thought or read or seen prepared me for that moment. As I stood looking down at the city 10 stories below, the train whistle spoke to me not only as the voice of the night but — suddenly — of Life: "I am here," it seemed to say, "in the dark, and I reveal myself only once. You know, now, that I exist. It is up to you to find me."

The future, the world — their voice was the same, blended into the tantalising sound of the train moving unseen across the summer night. It was that simple. I stood at a window and heard a challenge. I did not know that I would spend the rest of my life trying to make an appropriate reply.

The Pulse of the Night

During college I stood at night among trees — at loose ends, full of a vague, low-grade rage. At 22 I wanted much more than I could get. Nothing quite satisfied.

I stood there and listened to the tree frogs pulsing in the dark. They maintained a monotonous yet almost frantic pace that belied the blandness of the night, and the more I listened the more they seemed to match, to intensify my volatile inner state.

I waited for their tempo to change — to become more lulling, more casual: to reflect the timeless and calming rhythms of the earth: to be like wind in the pines, surf on the shore.

Not so.

The steady beat of the tree frogs seemed almost mechanical, manmade. Unvarying and mindless at two pulses a second, their sound remained fixed and meaningless, like the constant grin of an idiot. It was, I thought, like a hidden alarm system that had been triggered in the night. But no one was responsible for it; it would never be fixed.

Confession

I confess: I cannot remember a single term from college trigonometry; I always have to look up the difference between *discrete* and *discreet*, and I would probably blow up an engine if asked to jump start a car. Yet despite these and a dozen other incompetencies (— indeed, there is clear evidence that I ought to have my Adult's license revoked) on occasion I feel worthwhile.

Sometimes — and I say this cautiously, with the greatest possible deliberateness and care: sometimes when I write down words I feel as though they are God's own truth. They come from such an achingly aware part of me that they are like a sudden slash of lightning, a rustle of leaves before a storm, a bared body. They seem to be incontestable statements of the way things are. And for a brief while I feel myself to be on a high plateau where it would be possible for voices to say to me, serenely, from above: Yes, you have done it; you have perceived the nature of reality, as others have from time to time, and have rendered it faithfully.

Then, inevitably, such lofty moments fade, and there I am, with the dim wattage of Neanderthal Man flickering in my eyes as I try to follow the How-to-assemble instructions for a K-Mart barbeque grill.

Sunset Road Recalled

I wanted to be a father again, with children growing up in my house. I wanted to see them in the rooms of my house, and in the yards. I wanted to hear their voices among the voices of their friends on the front porch and in the back lots. I wanted them to ask me questions, and I wanted to answer. I wanted to be walking out of the garage with a sack of nails in my hand and I wanted to look up and see my son riding his bicycle along a nearby street: He would have come out of a neighbor's yard with his friends and they would be riding underneath the elms and he would see me there beside the garage and he would raise his hand in a greeting as he passed by. I wanted to see my daughter's face reflected in the mirror as she sat at her dresser, brushing her hair in the late afternoon light. I wanted to stand in the doorway of her room – her back to me, her brush moving slowly, her eyes looking at me from the mirror as she told me about the date she was having that night.

I wanted school books on the living room couch and a baseball glove underneath a bush. I wanted half-eaten crackers-and-cheese on a plate in the kitchen and a package of oboe reeds on the buffet. I wanted the telephone to ring: I wanted to answer it and then put the receiver down against the wood of the table and walk outside and tell my daughter, my son, that Sally, or Steve, was on the phone.

I wanted to be a father again the way a sheep wants to eat grass and a mockingbird wants to sing. I wanted to be in my natural place.

I wanted doors opening and slamming; I wanted giggles and shouts.

I wanted to be "Dad" once more, every hour of the day, each day of the week.

And I couldn't be. I was divorced. I lived apart, my children were growing and grown. They opened other doors than mine each afternoon; other people were in the house to greet them.

And when they said goodnight, I was no longer there to hear.

The Nature of Things

Mushroom eaters and psychedelics have often described the sudden perceptions they say they experience: the timelessness of a chair, the eternity of a candle flame. My own gropings into the nature of things have been undramatic, but from time to time I have also tried to express my attempts to grasp the essence of "other realities."

For example, someday I will manage to break the secret code of small back yards on Sunday afternoon: that elegantly shaded side of a wooden garage; the pale, sun-spotted grass beneath the nearby chinaberry tree: the intense and intimate presences that wait each Sunday behind an ordinary picket fence.

Or I will penetrate the mood of those small enclosed lots behind ranch houses when, at sundown, the chickens are almost ready to go up into the trees to roost: the place of eight or nine slender oaks that stand here and there above horehound stubble and caliche rocks: a private grove-and-pen behind the silo and the garden, a place where the hens move about casually in the shade like a pampered White Leghorn-harem idling away their last moments before flocking together out of sight to recline among divans.

If I am lucky, I might even understand the silence of a pecan grove in late-afternoon summer sun. It is a "terrific" silence, as Faulkner might have called it — "terrific" being his word to suggest what he could not actually express about the private, incommunicable "something" that an observer feels is inherent in objects and moments simply because they exist: the powerful "sense" of a thing — like a silence that is more than silence — which humans react to deeply but cannot describe because it is beyond the grasp of their perception and of language itself.

Lightening the Load

It was January, 1984, on a cold winter night in San Antonio. I closed off the rest of the house and sat by the kitchen stove at a small table. I took out a sheet of paper and divided it in half. On the top part I listed all the losses in my life during the previous four years. Then I drew a large X through each loss to make it stand out more dramatically. There were many X's. On the bottom half I listed the pluses of my life. There were three.

I stared at the two halves of the paper: the X's so overwhelmed the positives there was no doubt that at age 53 I had reached the bottom of my personal barrel. If I was going to survive I would have to erase the X's or they would erase me. They were like dead bodies I kept carrying around inside me — no wonder I was so heavy in spirit, so weighted down. I was not just depressed; I was suffocated, strangled.

Sitting by the stove I understood that for all useful purposes I was dead and would never come alive again until I let go of the past. So, on that January night, in a silent, cold house, I unloaded bodies. My arms folded, my eyes shut, I began to throw them out of my life — physically — with shakes of my head and grunts from my throat. People, places, situations — all that I had assumed was part of me, of my life's story, and wanted to keep: I threw them away. I purged myself in order to bring my mourning years to a close.

. . . You either do it this way — I told myself — or you will never again know the freshness of experience. You will be an embittered Scrooge — as you were today — looking out your window at children playing in their first snow: feeling nothing for the children, nothing for the snow — *feeling nothing at all.*

And I remembered, suddenly, a scene of the early 1960's: I was in Dallas, walking down a spring-haunted street near S.M.U. and I noticed an upstairs window in a house shaded by tall trees. The window was in a student's upstairs room like the one I had lived in as a college student in Austin ten years before. . . . I stood there on the Dallas sidewalk, on that sunshiny, leafy day, and it was the same day, once again, as my own sophomore year: an upstairs room on a side street, with music from a record player coming down through the trees, a college world nearby, life to be lived.

Sitting by the stove on that snow-silenced January night I lectured myself: If you don't stop mourning you will turn your back forever on that upstairs window with those white curtains blowing in and out. You will never know innocence again, and expectation. You will give up the pleasure of experiencing — simply and directly — the present.

I listened, heard what I had to say, and kept emptying myself like a wild sailor bailing from a leaky boat: Lighten the load. Throw out the losses. Live again with today's sun on your face. Forget the lost warmths of yesterday. . . .

PERSPECTIVES

West Texas

Space — West Texas space — is often misunderstood. Tourists traveling west see mesas and scrub brush and call what they see *empty*. They press on the gas pedal and hurry across a landscape that to them is too desolate and dry, too barren and forbidding.

To those of us who thrive on the desert land of West Texas, its openness is its glory and is not empty; it is filled with the spacious joining of earth and sky, and this constant blending of horizons deeply satisfies. We stand in the presence of immensities and are not overwhelmed. We are not lonely or bereft but feel curiously vital and at ease. We feel at home because the land frees us — allows us the luxury to be ourselves.

West Texas is dry but not lifeless, miles wide but not lacking in intimacies. It is a place of vistas and thus perspectives that lure one to search longer, look deeper; a place where the individual is not lost or submerged among countless others but stands out, makes a difference; a place where rock and shade and tree can provide a daily pleasure because surrounding spareness brings beauty more immediately into focus. It is a place where the sun cleanses us and makes us smile.

And the mountains: They are the heart and soul of desert country. They offer a source of strength and a sense of permanence. They jut up, stretch out, never die. They loom in the blue distance as visions and memories: ageless reminders of the grandeurs of life.

West Texas, then, has what was once called the frontier spirit: room to move around in, a lack of restrictions. It is a place inconsistent with pettiness or narrowness of mind. It provides, still, the freedom a stallion feels roaming over his grassland range: cloud shadows lying across a slope, a blaze of yellow among nearby trees, the sharp smell of greasewood lingering after a rain.

Walking

I started walking, purposefully, in college and have never stopped.

Walking, for me, began as an exploration of life represented by the streets of Austin in the early 1950's. It was the physical expression of the urge to know, to experience, and for four memorable years I experienced Austin on my two good feet.

Walking: it brings forth hidden riches, for a person seated is a person only in part.

Walking stirs up thoughts and sets them circulating. Step by step a walker loosens the framework of rigid ideas. A person walking is someone headed toward new possibilities and new considerations.

Walking is therapy. A walker allows tentative solutions to deep-seated problems to rise up from his shoes. Bubbles of insight gradually percolate through the blood stream and open new synapses, calm old fears.

Walking, of course, is exercise, and if a person cannot abide the idea of disappearing into gyms or "fitness centers" from 6-7 o'clock in the afternoon – just when the land begins to cool and shadows lengthen – in order to push up monstrous weights or enclose himself in a stark room and slam balls against a wall (classified in other societies and other times as harsh and unusual punishment), then walking in the countryside or down a tree-shaded, neighborhood street can be a way to tone the body as well as relax the mind. It is not a fierce, self-punishing act of muscling-up but, instead, an act of joining self-to-other – foot to sidewalk, skin to afternoon breeze, eye to yard or field – while swinging along at a personally satisfying pace.

Walking is reclaiming one's human inheritance. For thousands of years our ancestors, having no other choice, walked a lot, and their movements are still in our bones. To disregard such a legacy is to deny our past. Walking links us to the animal – the natural – world and allows us to connect ourselves again to ancient and satisfying rhythms.

Indeed, walking – when raised to its highest power, *wandering* – is a concept greatly revered in other parts of the world. In the East it is believed that wandering re-establishes the original harmony which once existed between humans and the universe. Said Brahma: "There is no happiness for one who does not travel." Said Buddha, as his last words to his disciples: "Walk on!"

Jack in the Box Heaven

Last night during the snow I sat smiling in a Jack in the Box. You see, I was sitting there in my old K-Mart winter jacket, listening to Muzak ("Flamingo," no less, with swelling strings calling forth visions of blue water and blazing sand) and looking at the young Jack in the Box waitress. She was shockingly childlike – hardly taller than a Double Meat with Cheese – and she was almost hidden by the counter: a fifth-grade looking girl with soft, wispy hair who nevertheless called out her orders with the aplomb of a Vassar finance major.

A countryish fellow came in, placed his order, and then told the waitress he had just got off work. He sort of announced the news to her, letting his words hang in the air as if hoping the girl would say, "Wow, just got off work, did you?" But – strictly business – she told him he was Number 12 – I was 11 – and he turned and took his seat. Rather forlornly he sat holding his reddish, chapped hands together on the table, his lips pursed, his career as a small-talk artist over for the moment.

I listened and watched and waited, and it was so good to be there in the booth – in my jacket, with it snowing outside, with my fellow human creatures of the earth nearby – that I could not help but smile a little. And the smiling felt good too. I was home again, for a while – home within myself. I had been away so long I wasn't sure if I would ever come back. But I had, and there I was in a Jack in the Box Heaven on a Saturday night – peaceful as a monk.

To the Man in Love Standing at a Pay Phone Outside Circle K

Earlier in the week, in your office, you were ostensibly you: you talked to your office staff – even laughed – as you had talked and laughed before. But in actuality you were vacant, unfocused: sunk in a depthless contemplation of Her.

She – this suddenly important Other – had become, incredibly, as real to you as your own face and body and thoughts. The awareness of her was an ache in the chest. "She and I" was a new dimension in your mind – shadowy in outline but electric in content. To live ("My God," you told me, "I'm 45!") was to be vulnerable again.

The world had tilted. Overnight it shrank – or expanded – and now contained, secretly, just two. The routines of daily life slid from view like a bundle of soiled shirts down a dumbwaiter.

Nothing mattered except passion.

. . . So there you stood on a Saturday afternoon, a man at a pay telephone with a quarter in your hand – making your secret call. Put simply, you were transfixed by the age-old magic of a man-with-a-maid. Your coin? A treasury. It did not give you her *hair,* or *smile,* but it bought you their promise. And as you stood there – your quarter deposited and a voice in your ear – life was reduced to another's breath.

An incredible chemistry was at work outside Circle K, for it had seemed a lifetime since you had said to yourself, "I want, I need."

You were *young* again – young as skin is, young as desire.

Listening to her laugh, you were content to let continents crumble into the sea.

I do not know what will come of this – a middle-aged man with quarters, with his maid. But I am certain that never again will you use such small change to buy the promise of such a shining happiness.

School Days

Student and teacher, I have been around schools for almost 50 of my 56 years. Chalk dust, not calcium, has shaped my bones.

I can't remember why, but I didn't start school when I was 6. Instead I went up the hill to Mrs. Baxter's house and learned to read in her small front room that smelled of dog. Mrs. Baxter, an old retired teacher, had spent most of her life holding thin Dick and Jane books in her liver-spotted hands.

When I was 7, I skipped first grade and began the second with Miss Breihan, who was young and pretty. I remember her lips (remarkably full and rosy), her voice (soothing and throaty), and her skin, very fair, with black, orderly hair lying in neat rows across her arms). I'm not sure what else I learned.

Third grade was Miss Sutton, who may have been 30 or 60: it is hard for children to estimate the age of adults. She always walked from the classroom door to her desk the way I saw (much later) the British actress Judith Anderson move about on stage: stately, controlled, austere. In my mind she is forever seated at her desk wearing a high-necked black dress and looking out at us with serene forbearance – her mouth curved slightly upward. She was my first experience with a Mona Lisa smile.

In the fourth grade Mrs. Moore wore rimless glasses that glinted and hid her eyes. She was big on making us cut out and color Thanksgiving turkeys. We put them up on the walls, and all through October, November, and December they marched wattle-to-tail feathers in a golden-brown frieze above the blackboards. (And believe me, blackboards were *black* in those days: none of your mod, pastel-green stuff back then. When we did arithmetic at the board the additions and subtractions looked properly severe: skeleton-white four's and eight's stood out against a black background as sober as the Depression.)

The fourth grade was when I also received a report card marked at the year's end – I'm amazed to recall the total – with 40 days of absences. I can still see that figure written in Mrs. Moore's small, elegant hand-writing. All winter I had croup and more croup and whatever else that wasn't croup but would nevertheless make my neck glands swell into egg-sized lumps.

(That was in the early 1940's in Kerrville, when doctors still made house calls. I would be propped up under a quilt in the bedroom, reading *Don Daring and the Temple of Doom* or a Hardy Boys or Ellery Queen, and I would hear my mother go to the front door. I would hear his familiar voice and familiar steps coming across the wooden living room floor; then the bedroom door would open and Dr. Knapp would stride in carrying his worn black bag: a tall, thin, balding man who never said very much but smiled as if he knew all the secrets: lips closed, the knowing mouth slit ear to ear.)

Mrs. Butt – yes, it was indeed her name and we all came to terms with it – had graying hair and was jolly and sociable enough for us fifth graders. We were getting to be old hands now at the business of elementary school, and Mrs. Butt was the first teacher who sort of – not entirely, but sort of – seemed as if she could be a friend – as if she were a person as well as a teacher. She smoked, I think, and might even have liked martinis. Who knows the worldliness a teacher is capable of if she actually *smiles.* . . .

The next year was the year the high school/junior high had to be remodeled because of the fire (we only lived a block away from the school grounds and on that September night we stood on our front porch and watched the second-story biology room erupting smoke) and thus our sixth grade did not move, as was planned, from the elementary building. We suffered loudly for a while at having to remain there with fourth and fifth graders instead of moving across the tennis courts into the Big Building, but we survived. Sixth grade was fine except it brought the beginning of Word Problems in arithmetic ("If a boat is going from the north side of a river to the south side at ten miles an hour, and the river current is six miles an hour, how long will it take the boat to reach the other side if an east wind is blowing at 12 miles an hour and the price of corn is. . . ." – *that* kind of problem). It was my first real time of agony in a classroom. I learned what it was like to sit at a desk and think and think with a mind that had slowly turned into banana pudding.

Sixth grade was the year of marathon Monopoly games. An epic one began one Friday afternoon after school at my house and continued through supper until bedtime. Then at 7:30 the next morning — cold, wet January Saturday morning — Sam, Glenn, and Billy Ed were waiting on the porch when my father went to the door. We played in the front room by the roaring fireplace until noon.

That was also the year of my Amnesic Home Run. It happened like this: At ten o'clock recess we were playing baseball — softball, that is — on the playground. I was second baseman and at one point when a runner came my way his knee socked me a good one on the cheek bone. Apparently I blacked out in some way — had no recollection, afterwards, of the next three hours — but still continued to function in a kind of addled state. Indeed, so the legend goes, my next time at bat I hit a home run — the only one I ever hit in my life. . . . I went on back to Miss Real's room until noon and walked home to eat lunch as usual — but by then sporting a hefty knot instead of a cheek bone. I stayed home that afternoon with an ice pack on my face, trying to deal with the strangeness of the lost three hours. (The Good Doctor Knapp counseled my frantic mother over the phone that all would be well.)

After school the guys rushed over, and I basked in their attention as I lounged on the front porch swing with my ice pack. They stood in their semicircle, asking the same questions over and over: "You don't *remember* when Marianne knocked over the waste basket? You don't *remember* when Miss Real gave us our tests back?"

Seventh grade was roller skating, spelling bees, Halloween carnivals, girls getting taller — pretty standard stuff. But there was, for me, a minor moment of infamy.

You see, I was beginning to be fairly conscious of words and their effects. When we had a contest to name our junior high newspaper, my suggested name — I think it was a rouser like "Junior Hilights" — was chosen. Anyway, you understand my point: I liked words and liked learning new ones.

On my day of shame I had gone home at noon, eaten lunch, and apparently had picked up something to read before going back to school. My class after lunch was art with nice, easy-going Miss Steffens. . . . I was at a table working on my project — carving an eagle out of Ivory soap – when Miss Steffens went to the board to write some instructions for us to copy. She was standing there, writing in her graceful script, when I . . . when I what? wanted to impress her as a budding intellectual by showing off the new word I had just read at noon? Probably. Anyway, I *thought* I was asking in a polite if flippant way if she would move her *back*, her *backside, posterior.* At least, that was my intention when I sang out in the quietness of that lovely hill country afternoon: "Miss Steffens, would you mind moving your rectum?"

I wish "Candid Camera" could have preserved on film the expression on Miss Steffens' face as she turned around and looked at me. I doubt that I was paying any further attention to her. Having delivered myself of my impressively bookish word, I assume I bent back down to resume carving on my Ivory eagle. It is to the everlasting credit of Miss Steffens that she – that wise woman – perceived in me a lack of malice, of smart aleckiness, in my question and thus did not choose to humiliate me. She let the moment pass and moved to one side.

Later, of course, as days became years and I understood the depth of my blunder, I had a hard time making eye contact with Miss Steffens in the hall.

Music in My Life

Young people who listen to current musical groups such as "Metallica" and "Guns N' Roses" tend to regard music of the past – popular, swing, jazz – as just so much Elevator Music: hopelessly and forever Lawrence Welk.

So it goes. Music will always be as divisive as politics and religion: to True Believers there is only the musically Right (sublime) and the musically Wrong (ridiculous). But just as steak and fried chicken do not go out of style – or seashores, pine trees, or backyard dogs – neither do Duke Ellington and Peggy Lee.

I confess: musically, I am forever a child of the '30s, '40s, and early '50s – when Lyric was king, Melody his queen, and Sentiment their faithful retainer. My life is encoded in the recordings made during those years, and whenever I hear them I relive – for 2½ minutes or so – the times and places of my early loves and longings, innocence and youth, enthusiasms and despairs:

"Lost April," Nat King Cole. On April nights when I was a boy my family and I would say goodbye to uncles and aunts and cousins as they got into their cars at the front gate. Dogs barked in the neighborhood and a light breeze carried the pleasantly bitter smell of damp oak leaves. Behind us, the dim porch bulb shone through the low tree limbs; shadows were spread across the yard. The voices of the relatives rose, then were half-lost in the night. There were brief, fading laughs. Finally the cars began moving down the gravel street, arms still waving out of windows, heads silhouetted. Only when the cars finally rounded the corner by the elementary school a block away did we turn and start walking up the curved sidewalk toward the ever-abiding porch and house.

In a way I will always be there – on that sidewalk, under the trees. The pale half-light of the porch bulb stays in my memory like a pilot light: a persistent glow that keeps shining through to me, giving a kind of guidance and assurance from my past.

"Jingle, Jangle, Jingle," Kay Kyser. When I was 11 years old my family bought a small record player. However, since our home town did not have a music store, I had to wait until the family made one of its semiannual trips to San Antonio, 60 miles away, before I could add a few records to our almost non-existent collection. I went to the basement of the Kress store in downtown San Antonio and stood in front of the long rows of singles and albums: Columbia (red label), Victor (black label), and Decca (bluish-purple label), all 78's and all capable of turning into wavy waffles of carbon if left in the back seat of our car in the sun or else having big half-moon cracks appear in them if I picked them up carelessly. My first buys were *"Alla en el Rancho Grande"* by Gene Autry, "The Hutsut Song" by the Merry Macs, and "Jingle, Jangle, Jingle" by Kay Kyser. Carrying them upstairs and out the Kress door onto the sidewalk of Houston Street, I felt as worldly as Cary Grant.

Once home, I would put a record on our record player and listen to it perhaps a dozen times straight through, hardly letting it end before picking up the player arm and setting the needle back on the first groove: glorying in the fact that I was listening to one of the nation's new records just as people were doing in San Francisco or Philadelphia. My father was at work at his feed store, my mother was hanging out laundry on the back clothesline – family life was going right along – but I was in the living room listening to Harry Babbitt and Trudi Irwin sing – over and over – the very song I had heard them sing two nights before on the radio: "I got spurs . . . that jingle, jangle, jingle . . . as I go riding merrily along. . . ."

"Dance, Ballerina, Dance," Vaughan Monroe. It is 1945 and I am 14 years old. It is a cold winter night. The radio is on in the living room – where Daddy is reading the *San Antonio Light* by the fireplace – and I am seated at the kitchen table after supper, working on a project for school. (Saturday night: the sound of my mother ironing shirts in the bedroom, the weak yellow glare of the kitchen light bulb, the heat and security of the kerosene stove in the corner, and the baritone richness of Vaughan Monroe filling the house – bringing to us in our small town the glamour of New York and all the worlds beyond it.)

"Peg O' My Heart," Buddy Clark. At the 1949 senior prom couples are dancing on the smooth cement floor of the roofless pavilion. Colored lights are strung through the limbs of the surrounding trees, and beneath the balcony – below the wooded hillside – the Guadalupe River shines in May moonlight.

I am dancing with Anne Smithson, who wears the orchid corsage I had pinned on her strapless white dress hours before (in her living room, with her mother watching). She is still radiating "White Shoulders" perfume – the most overwhelming, the most devastating, the most romantic scent I have ever experienced. She smiles at me as she dances, showing her shiny, now-perfect teeth (– the braces she had worn since elementary school having finally been removed during her sophomore year). She is 17; she is beautiful. She has milky, bare shoulders and a Valentine-shaped face. (We will continue to date for the next several months – until she goes off to college and meets in her piano class the music major she will marry and live with for the next 35 years.)

From the jukebox loudspeaker attached to a pole Buddy Clark sings in his pleasing Broadway baritone. Full-bodied, accompanying strings swell behind his voice, lifting it above us — above all the decorous, sweating, high school bodies — into the warm central Texas night.

"Again," Gordon Jenkins: that's on the jukebox in 1949. The guys and I are sitting around in a cafe the week after high school graduation. We are drinking Nehi oranges and Pepsis and looking out at the passing cars, wondering what *college* means as we idle away the afternoon. . . . "Again . . . this couldn't happen . . . again." Well, high school will certainly not happen "again." What will ever happen to us — "again?" The massed chorus on the jukebox swells, the record ends. Myrna Colbath, the middle-aged waitress, moves about slowly emptying ashtrays and wiping tables with her damp rag. She doesn't look at us and she does not seem to be listening to the jukebox. She acts as if she couldn't ever be bothered — "again" — by anything: by either sentiment or song.

"Concerto to End all Concertos," Stan Kenton. On an afternoon in spring, 1950, I walk across a college campus and hear unisoned trumpets in a small wooden building set off among shaded sidwalks and hovering trees. The college dance band is practicing Kenton tunes. Perhaps five or six trumpets — full-bodied, heraldic, sustained — bend themselves into strips of gold-metal sounds. They seem to catch on their surfaces the last glowing fires of the April sun. . . . I stop, listen, move on — continuing another of the aimless walks of my college years. As I walk, the steady trumpet sound bores along in front of me into the very air, carrying across the campus and the late afternoon the sad music of spring — perhaps, I think then, at 19, the music of life itself.

"All the Way," Frank Sinatra. I am 27 and driving along a coastal road near Corpus Christi with the car radio on. I am brooding about events of the previous week when I had broken up with a young woman. (. . . I had bought the ring, she had told her friends and relatives about our wedding-to-be; we had given an announcement to the weekly newspaper in the small town where we were teaching. Yet as I had walked out of a drugstore in the late afternoon and unwrapped a stick of gum, I had glanced across the street toward an ordinary-looking house with ordinary porch and windows and yard — and I had asked myself: Do I want to walk — right now — into such a house somewhere and shut the door and begin living a married life, just Mary and I, forever and ever, amen? My stomach, tying itself immediately into a knot, gave me the answer.)

Fraternity

Ordinarily I don't present a very pleasing sight either to truck drivers or waitresses. My glasses, my serious-looking academic air usually make them ill at ease. They feel that somehow I am going to end up Acting Smart. But on a trip to Houston recently I had the curious experience of being dealt with in *too* friendly a manner — treated *too* courteously.

Twice within an hour I stopped in small farmland towns — towns with large numbers of blacks and Mexican-Americans standing around on the hot noonday streets — and to at least one waitress and one truck driver I seemed to be the lone unthreatening figure in an otherwise threatening landscape. In Seguin I placed an order at a Dairy Queen right after a graying Mexican-American man in khakis had placed his. Although I noticed that the sullen young Anglo waitress had taken the man's order and then brusquely turned away, I assumed at first that her action was innocent enough. Almost anyone asking for service on a sweltering August day can expect to have his order filled perfunctorily, even rudely.

Yet boredom and summer heat were not at the heart of the matter. When the girl returned with the man's hamburger she purposely held it back from the window so the man was forced to reach for it; then, after he had taken the hamburger and withdrawn his arm, she carefully, solicitously — and quite pointedly — pushed my own Coke and cheeseburger through the opening and turned the napkin holder conveniently my way. And as she handed me my change the subtle but unmistakable *something* in her voice formed a silent declaration that I could hear quite plainly: "We take care of our own around here. . . ."

The experience was repeated later outside a small roadside cafe – in the heart of Black Country just past Flatonia – when a truck driver almost swamped me with well-wishing and neighborly concern. He was sucking on a toothpick and standing outside his rig when I stopped, got out, asked if the cafe across the highway served beer: Flatonia was in a dry area. He took out his toothpick and shook his head no. Then, almost sorrowfully, as if deeply sympathetic to my need, he told me how to find a package store just a little way down the road. For a moment I thought he was going to get down on his knees and draw me a map in the dust.

I drove on, embarrassed at these overly enthusiastic greetings by people who seemed to move around so tensely in a world filled with alien presences. Their friendliness was like a coded message of solidarity from the Fraternal Order of White Skins.

Rejection

A man can be rejected by his friends, relatives, employers – perhaps even by his children, his country – and still retain a saving sense of identity. However, let him be rejected by a woman and he suddenly feels he is *nothing.* The pain is so acute, so overwhelming that his chest squeezes tight, he cannot breathe. He is miserable.

Look at him: He is standing in his house, at a window, and he is thinking of the Other Man beside Her on the couch, his head in her lap, her arm around his shoulders. He thinks of the man kissing her, touching her flesh – and of her then sinking against the newcomer the way she had sunk against him time after time. He thinks of them that way and. . . .

Well, if he is like most men, he will survive. He will absorb his pain even as he continues to suffer – in silence, or in tantrums of jealousy, or in drinking: in predictable bouts of gloomy behavior. Ultimately, however, he will manage to recover his equilibrium and get on with his life.

But there are exceptions: those childish, dangerous men with eggshell egos who not only cannot accept rejection but who become deranged by their sense of loss. They end up committing the unconscionable acts that cause readers of newspapers to shake their heads in disbelief: men who kill ex-wives or lovers and then, unable to tolerate the horror they have wrought, kill themselves.

I have often wondered about the psychological make-up of such a man. I assume that he is driven to the point that he must try to deny – rhythmically, blow after blow – the thought that the woman he considered *his* will no longer be intimate with him and him alone. He feels he must fight back against such an unacceptable thought by striking out against the woman instead of himself. His repeated strokes are frantic acts of denial – each blow saying no, no, no to his own desparate emptiness. Violence becomes the only way he can still reach her, continue to be next to her – occupy again the same physical space they once had shared.

(There are blows because there are no longer kisses. His eyes cannot linger on her face, her body, her clothing; she has replaced smiles and pleasantries with argumentative, distancing words. She has looked at him as if he were a stranger, an enemy – and he cannot stand it.)

I think violence would be such a man's frantic, wordless accusation: "You've ruined it! The good thing we had together – you threw it away. . . . you're even glad it's gone! I want to hate you now, not only for destroying what we had but also because when you pulled away you made me realize how much I needed you. I want to punish you because by leaving me you have made me feel weak and helpless and alone while you stand there, confident and secure – with that *other* who has taken my place."

This crazed and sickened man – no longer the law-abiding father, the churchgoer, neighbor – feels he has lost everything, has reached the end of his rope. Freed from all normal human restraints, he crosses beyond the last barrier of civilized behavior. Possessiveness and hate, violence and self-pity – all become one in the destructive surge of his final act.

Truth

Buzzards were circling in the seven o'clock sky as I drove down the West Texas highway. I still had a long way to go and Sunday night was closing in on me.

I dialed idly past a few stations on the radio until I began listening to a Baptist church service in Lubbock. As I listened – and as I continued to look out the window – the radio and the San Angelo-Big Spring land began a strange little tug of war for my allegiance: Whose side are you on? each seemed to ask. . . . Listen, look, know the Truth.

"Thee . . . Thou" – said the voices from Lubbock. "Our Father . . . Jesus." The sound of baritones and massed sopranos filled the air.

Miles of greasewood. More red cows eating. More shimmering green fields.

Prayers and preaching, the Holy Spirit . . . *churchliness.* Ushers and hymnals. Red ribbons pinned on visitors.

Goats, windmills, knolls. The leisured ending of a long afternoon.

"I love to tell the sto-o-ry . . . of Jesus and His glo-o-ry."

The sun dying impersonally in the wide white west.

"I was sinking deep in sin. . . ."

Sin – that puzzling word. I looked around, trying to catch a glimpse of something sinful along the highway.

"Love lifted me (even me) . . . Love lifted me (even me)." . . . "And Jesus sayud, 'I am the Resurrection and the Life.' . . ." Jesus with a Lubbock accent. . . . "You ask me how I know He lives; He lives . . . within . . . my harrrt."

Mild green mesquites leaning almost neighborly over a fence at sundown. (But they do not know Jesus as their Savior.)

"He walks with me and He talks with me."

Night comes, envelops my car.

The preacher in Lubbock invites us now to come forward. His voice is full of ministerial quiver and power.

The night, huge and silent, stretches toward Saturn.

Young Conan

My son has become a bodybuilder. At 21 he is shaping his pecs and lats (pectorals and laterals, if you don't subscribe to *Flex* or *Muscle and Fitness*) and is following a strict, specialized diet in order to become ripped (his body free of fat). He is not interested in merely looking good in his T-shirt at the Pizza Hut nor is he just another of the many guys who occasionally "lift." He is serious. He has become a student of nutrition and super abdominals in a way he was never a student of sonnets and isosceles triangles.

It's a tricky business being the father of a dedicated bodybuilder. I want him to do his best, but I have reservations and concerns.

We have had, of course, our father-and-son talks about steroids, and I have come away from them satisfied that he is taking a sensible approach. He acknowledges their effectiveness: they do create sensational "mass," they are widely used by athletes who want to become "the best they can be," and yes, guys do use steroids at the gym where he trains. But he is not interested in taking them, he tells me, because they are a short-cut means to the end he has in mind.

I repeat, he is serious.

. . . Okay, I'll just go ahead and be out front: I hate to see my nice young-man of a son distort his body. That's my main worry, I guess. I've seen weightlifters and bodybuilders going into gyms, and — those shoulders! those thighs! Their muscles are like bulging sausages ready to explode. They are, in my view, young men marching to a disturbing drummer.

Drummer: I look at the word and I remember other enthusiasms, other years. . . .

When Byron was a freshman in high school he played the timtoms — three drums hung in a cluster from his neck. I sat in the bleachers at football games and watched his band march out smartly at half time in their blue-and-white uniforms. They were good and he was good. I listened as the sounds of those triple-drums of his echoed against the stadium walls and his band went through their snappy routines, and chillbumps climbed up my neck.

As a senior in high school he gave up band to run cross-country. He ran at school, he ran on weekends, he ran through parks and up hills; he ran in the first light of dawn and in the bare light of dusk. On Saturday mornings at district meets I watched him take off in a blur of other runners wearing their school colors – watched his smooth stride and his blond hair flopping, yelled him on toward the finish line.

During his first year in college he was "into biking," and I helped him buy his first Schwinn racing bike with its skinny tires and gleaming chrome. He rode down back highways – 20, 40, 50 miles: head bent low in the helmet, legs pumping, blue racing shorts skin-tight against his thighs. He rode to his summer job as a life guard; he rode as if riding were the answer to all his questions.

Then, in his sophomore year, he began to frequent the gym, to focus on sets, on abs and glutes. He talked about his training partner more than he did world history; he concentrated on fat-free foods. He had become a bodybuilder.

I don't know, really, if I should worry about him. I have always accepted and encouraged his other choices: his way of achieving, of pushing himself. What he is doing is undeniably healthy, and the dietary regimen is certainly commendable – the egg whites, the chicken breasts, the shark filets and tuna fish.

It's just – troublesome. I don't know where his body-emphasis is leading this time. I am the father of the son I love – the son with biceps like rocks, the son who wants to be "cut to the bone" – and I want him to be okay, that's all.

Dear Ann Landers: Does my son have a problem or do I?

Flood

It started raining at midnight on July 16. I stood in the dark hunting cabin and watched the surrounding trees flare like dim phosphorus in the constant sheet lightning. It was as if the hill country had suddenly become Dr. Frankenstein's laboratory: thunder cracked and exploded overhead, the lightning kept up its strobe-like flickering, rain came down on the cabin's tin roof like a sky-bully determined to beat up the earth. I would try to lie down on my sleeping bag and just listen to the rain, but every half hour or so I would go back to the window. I had to witness such an awesome display of power.

At three-thirty, when the storm passed on, I opened the cabin door and went outside. The night was subdued except for the roar of gullies and dry creek beds as the water poured down from nearby slopes and into the lower pastures.

The next morning I checked the rain gauge on the fence — it showed five inches — and turned on my radio to the Kerrville station. There was bound to be a disaster somewhere in the hill country, I thought — there always was in the summertime when it rained so torrentially during the night. Tourists in Kerrville and nearby towns would try to cross low-water bridges when they shouldn't — and their cars would be swept away. Summer camps along the Guadalupe River would become marooned — perhaps threatened — by rising water. Families who had gone to bed the night before in their pleasant river-front homes would awaken to swirling water in their living rooms.

Sure enough, a tragedy had already occurred — at Comfort, southeast of Kerrville. The first radio reports at 9:00 were lacking details, but apparently a bus loaded with church-camp teenagers had overturned trying to cross the Guadalupe. Some of the children were safe, some had been swept away, others were still clinging to cypress trees along the river. Rescue helicopters from San Antonio were on their way to the scene.

News updates during the day clarified the earlier reports. It was finally established that a bus and a van evacuating 43 teenagers and adults from the Pot o' Gold Baptist church camp had not tried to cross the river, as was first thought, but had gone down a back road in order to get to higher ground before the river flooded the camp. Four previous buses had successfully turned onto the road, but the Pot o' Gold bus stalled and the river spread around it. The campers got out of the bus — joining hands to form a chain in the water — but the surging river began to carry them away. The bus overturned, floated past them, and sank. The campers in the van got out while it was still floating; then it too sank.

Each day thereafter I would go to town, buy a newspaper, and read the accounts by survivors telling about the terrors they had faced. I kept up with the efforts of rescue groups trying to find the bodies still missing in the river. I also read a weather analyst's description of what had caused the flooding: the storm system that passed across my hunting cabin at midnight had stalled, perversely, near the little tourist town of Hunt – where the north and south forks of the Guadalupe join – and dumped 11 inches of rain before daybreak. By 7:30 on that gray Thursday morning the full force of the 11-inch deluge was being felt at Comfort, 30 miles downstream.

As I read the newspapers in Kerrville and then drove back to the cabin, images kept haunting me:

Here were ordinary kids who had come down from a hot Dallas suburb for a week in the pleasant river-world of the hill country: a week of swimming, singing, Bible-reading – church-camp delights. Then barely past daybreak they found themselves sitting on a bus, staring at a strange new river. In those first moments, however, life was still okay and they were okay: Nature was acting up but adults were still in charge. The morning, though heightened and dramatic, was a morning like other mornings and they sat as they had always sat on buses that took them to school, church, sports events. They bounced along, looking passively at the back of the bus driver's head, and they watched the other four buses in front of them begin to make their turn uphill. . . . It was all still understandable, you see. It still made sense. Soon they would be wherever it was the adults thought was safe, and then later in the day they would be calling home to tell their parents and friends about the wild experience they had just gone through.

At this moment as they sat there in their blue jeans and T-shirts the surge of Hunt floodwater caught their stalled bus, and the screaming started. Nothing for any of them would be the same again.

(How do you become selfless – a hero? You don't go into training. Suddenly, the moment to act is just *there*, and you respond – or you don't. The stadium is collapsing, the snowslide begins, the rains come and the river rises. . . .)

Mike Lane, 18, taught Bible classes at the Seagoville Road Baptist Church outside Dallas. He could have made it from the bus to safety and lived to read about his narrow escape. But he didn't. He helped younger children make their way to dry land — pushing down part of a barbed-wire fence, holding it there while the smaller campers jumped across. They made it. He turned back to help others, grabbed a log, tried to paddle it. He was swept under and out of sight.

Two boys clinging to a cypress tree began to ask each other: Do you think we're going to die? There had been no time to discuss, leisurely, life and death. At 7:45 the terrible question was starkly before them. All that had been stable and predictable in their lives was rushing past them. Nothing — not Bible-reading, not being good to others, being helpful — had prepared them for such a blunt, existential moment. They clutched the slippery trunk of the cypress as uprooted trees, fence posts, boat docks churned by them.

One of the boys was John Bankston, a 5'1" 17-year-old from Balch Springs who pumped iron for hours each day. Called Muscle Man by awed Pot o' Gold Campers, he had a dream of going to Baylor and playing football. . . . When the bus had stalled and John saw that a friend with a cast on his leg was having a hard time getting out, John lifted him up and carried him on his shoulders. The water was waist deep, then chest deep, then deeper, but John and his friend with the cast made it to the cypress. There, in the water, John held to the tree and the friend held to John's shoulders — for the next hour and half. They talked and they waited. Finally the friend floated down to another tree, where he was later rescued. John, exhausted, disappeared into the Guadalupe.

Thirty-three campers and adults were rescued. Eight drowned. One girl fell to her death from a helicopter rescue rope. All bodies of the victims except one were recovered during the week after the flood.

John Bankston's body has never been found.

ANAIS: A STORY

I.

A day seldom passes that I do not think of Anais Brautigan. She is dead now – her ashes in a jar – but I cannot leave her alone. She haunts me dead as she haunted me alive, and I have a need to tell about her: to reach back and re-examine the summer we were together on River Road and the time that followed.

It was May and I was at loose ends, recovering from a divorce after 18 years of marriage. A friend said, "Hey, why don't you look up Anais. She's . . . *different*, and terribly bright. She's had more things happen to her than the law allows, but she paints and writes poetry and raises greyhounds – I bet you'd like her."

I was teaching then in San Antonio, and with school nearly out – and a bleak-looking summer ahead – I thought, okay, why not, I've got nothing to lose. So I called her up one Sunday. We talked a bit and Anais told me how to get to her place south of town.

I followed her directions and drove through San Antonio to the city limits. I turned off the interstate onto a paved road that went past heat-seared fields, roadside junk piles, occasional oil rigs pumping among mesquites. I bounced onto a dusty, unpaved road and kept going past yellowing gourd vines and cactuses. It was depressing flatland under a hot South Texas sky.

When the road came to a T, I saw a small handwritten sign on a board – River Road, No Trespassing – and turned right toward a line of tall trees. I drove through a natural archway made by two huge oaks, entered a shadowy tunnel of druidic-looking elms, and there it was: a brief clearing in the river-bottom jungle and a small white frame house within a new chainlink fence: Anais Brautigan's hideaway acre.

I drove up to her gate, and before I had time to wonder if I should honk she was coming out the front door. She walked towards the gate, smiling, smoking a cigarette and holding a smooth black greyhound on a leash. She wore dark glasses, tight blue-jean short-shorts, a man's white dress hat, and high heels.

"Do-you-have-your-toothbrush?" she asked in a Germanic drawl, still smiling, looking at me above her dark glasses as I got out of the car.

And that was that: The next weekend I moved in with Anais, her greyhounds, paintings, radio tuned 24 hours a day to low-volume Beethoven and Pavarotti, and I stayed until the end of August.

I must be clear about this: I felt no love for Anais in the beginning. I don't fall in love easily. But there was an immediate companionship between us — an automatic ease. When we started talking that Sunday afternoon it was as if we were picking up a conversation we had stopped temporarily a few days before.

Those first summer days: they were like a ball of string that just kept casually unwinding. We talked, and ate, and drank coffee; we made love, and read, and did chores around the place. We sat in the side yard under the tall, interlocking elms and drank beer and talked and then made love and held each other in the dark. Another day would begin and our routine would start all over again.

Sometimes in the afternoon I would go for a walk or a run down the road. Sometimes — rarely — a friend of hers stopped by. And sometimes we would drive in to a shopping area at the south edge of town. We bounced along, the two of us in her truck, past the dust-covered hackberries standing in the blazing heat, and we were in our exact, proper places: as if we had spent the first fifty years of our lives getting ready to sit together in a Mazda cab at 3:30 on a summer afternoon. We would go into a liquor store next to the Value Club and walk around among the bottles. Anais would get a fifth of Jack Daniels and a carton of Marlboros; I would get a bottle of white wine. As we stood together at the check-out counter I would look up into the small TV screen of the surveillance camera and see her there beside me: the curiously diminished Anais in her small gangster hat smoking her small cigarette (— as though already miniaturized into memory, into a photograph). We would get back into the truck, drive the lackluster southside roads past the drab subdivision houses, go back under the interstate — and it was as if we had been off on an extended vacation: the connection between us was that calm, that close. It felt like a married-life homecoming to be heading toward Anais' little house with the trumpet vines growing across the west wall and circling down over the butane tank; with the two greyhounds asleep inside the house on the front room rug; with the summer sun bearing down and the rusty air-conditioner roaring in the bedroom window as we pulled up next to the gate.

(When I woke on the first Sunday in June the windows were open and Anais was curled next to me, still asleep. I lay there a while, my arm beneath her waist, my face touching her hair. Then I raised up carefully and sat against the headboard with a pillow behind my back.

It was 8:30 and the house was still. I looked at the shadowed corners of the room, the wood stove, the clothes hanging across the tops of doors. I looked through the window screen at the trunks of the elms. I heard the doves at the river. It was like the first morning of the world, and I was content.

I sat there in the bed and thought of the Sundays of my adult life – the days that shriveled my spirit, the days I always had to kill, hour by vacant hour – and then I looked down at the evenly rising and falling shoulders of Anais and smiled: It would be a River Road Sunday all day long.)

There were only the two greyhounds that summer: Cleopatra, the regal black queen mother, and Apollo, the tawny son that Anais planned to race in the fall. I would stand in the front room, drinking coffee, and look at River Road royalty stretched out in their chairs. They were dozing or sleeping – who could tell. Their chins were slung off the edge of the two big chairs almost to the floor. If I had fired a cannon over their heads they would not have moved an eyelid.

One afternoon when Anais and I were giving them a little exercise in the back lot, I decided to sprint ahead toward the river and invited them to join me. "Come on!" I yelled over my shoulder to Apollo, and he took off toward me like a shot.

I didn't know anything about greyhouds – gazehounds, as they are called. I didn't know that they do not swerve to avoid humans as other dogs do. They keep coming, gazing ahead, gazing beyond. . . . Apollo did just that. He ran into me and across me and kept on going. It was while I was on my knees in the dirt, with Apollo blazing toward the river in his incredible rocket lope, that I remembered the story Anais had told me: A farm-friend had come visiting one afternoon and had yelled, "Oh, Apollo! . . . Cleopatra!" as I had done, motioning to them from below the garden. They had shot toward her, doubling, tripling their speed in an instant, and while she stood there in the field — this kindly, smiling, white-haired grandmother — Apollo had flattened her like a white-tasseled stalk of corn.

Several mornings we meandered through the county in her truck. We listened to her tapes — Wailin' Waylon, Madonna, Kitaro — and drove past junky trailer houses with half a dozen cars pulled nose-in toward the trailer door, past faded curve-of-the-road beer joints and abandoned old houses half-visible among vines and weeds and drooping tree limbs. Sometimes we would stop and make our way past the dumped debris of old sofas with the stuffing coming out and pick bucketsful of wild grapes that grew along the fences. Or we would end up at Braunig Lake and Anais would get out her paints and do a few watercolors while I wandered along the shore. At dusk the wind would get up and the little waves would come washing in. Men in worn tennis shoes and with toothless, sunken jaws would stand next to their tents, beer in hand, and tell us about the red fish they would be catching after dark.

On the way back from the lake one night we started talking about cornbread — good, satisfying, red-beans-and-butter cornbread: family recipe, flaky, made-with-buttermilk cornbread. We worked up an enthusiasm and an appetite for cornbread we did not know we had. I pulled off the highway at the next stop-and-shop, bought buttermilk, and for supper Anais made a panful of cornbread that we ate like cake.

(I don't think Anais ever hurried in her life. She had an inner metronome that measured her stride, and she moved the way a cat would move walking upright. It was both a physical mannerism and a philosophical statement. I am in control, the walk said. I do not plan on being seriously deterred or surprised or thwarted, but, if I should be, it does not matter.

It was good to watch her anywhere — drifting along the aisles of a supermarket, looking for a three-way electrical plug or a can of mushrooms, or coming out the front door of her house to greet a visitor: With the screen door shutting gently in her wake, she would pass through the affectionate front-yard lunges of the greyhounds like Faulkner's barefooted boy moving obliviously through the maelstrom of hurtling, wild-eyed spotted horses.

I liked it best, though, when she was prowling the side of a pool table when we went over to the Friendly Tavern. With her black-and-white gangster hat pulled low and at an angle, with her constant cigarette at its downward slant, she circled the tables in the subdued bar light. She would chalk her cue stick and lift her gaze slightly to look over at me above her dark glasses, then look back down, slide a few steps along the table in her narrow black boots, make her shot, keep on moving in her cat-like way — already preparing for her next shot as the ball rattled down through the chute.)

One night she fixed rainbow trout with brown rice and white wine. She put on a dress and earrings and high heels. The light was dim, just a candle on the kitchen table. After supper we sat outside in the dark, talking. The trees loomed about us and from time to time a strange, hidden bird at the river — almost like a manic jungle-bird in a movie — gave out its screaming teakettle calls. We talked and listened to the bird and finished off the bottle of wine.

Anais took the wine glasses inside; I went around to the back of the house to bring in the laundry from the line. She took off her shoes and went back outside to water her marigolds by the front porch. I had an armful of shirts and was reaching for the socks when I heard her yell out my name. I ran inside and found Anais holding on to a chair in the front room, her bare foot raised. "Something bit me," she said. She had turned off the hose and started toward the porch, she said, when she felt a stab — sharp and hard.

I made her sit down, and after I tilted the lamp for better light examined her foot. I couldn't see the sign of a sting or any swelling. There didn't seem to be any mark at all. I told her it was probably a scorpion, but she was shaking her head no: she had stepped on scorpions before. This was different. I said, well, maybe it was a tree asp. They were around and they could hurt.

Anais was one to downplay discomforts and pains – almost to belittle them as being beneath her – but it was obvious she was hurting. "I need to lie down," she said, and I helped her into the bedroom.

She kept getting worse.

"I think . . . it might have been a water moccasin," she said. "Maybe a little one." At first I didn't think that was a possibility – the house was 50 yards from the river. But she said – her voice barely audible – water moccasins were known to breed that far from water. She thought maybe they had made a den under the house.

I called Southwest Baptist Hospital and told them we were coming.

"Let's go," I said, and I half-supported, half-carried her to my car.

As I was driving fiercely through the southside San Antonio streets, Anais asked me a for a piece of paper and I tore a page from a little notebook I carried in my shirt pocket. I turned on the dome light and she began to write as she leaned against her door. "What are you doing?" I asked as we hurtled along. "Writing my will. . . . I don't think I'm going to make it to the hospital." She wrote her holistic will, as she called it, on the little sheet – stating in one sentence that she was leaving all her earthly goods to me. She signed it, handed it to me, and sagged further against the door.

At the hospital the doctor on duty gave her a shot of Benedril as she lay on an emergency room bed. He could find no fang marks, he said, and he could not explain her symptoms. Maybe a snake bite; maybe not. After we waited another hour or so Anais said she felt she could make the drive back home.

She remained sick the rest of the week. She would get up for a while but would soon go back to bed. Her ankle was swollen a bit but not too much. She kept running a degree of fever, off and on, and had a general sense of malaise.

One night I heard Apollo barking out in front. I turned on the porch light and went outside. About a yard beyond the porch, in a coil, was a small, grayish-green snake. I yelled at Apollo to get away and carefully reached for the hoe that was leaning against an elm tree. I struck the snake a good lick and killed it. Then I called through the bedroom window to Anais. She came to the front door and I showed her the snake draped over the hoe. "Well, that's one," she drawled. "The brothers are waiting for you under the house, I suspect."

The next morning I got a flashlight and peered through an opening in the tin bottom that ran along the side of the house. There certainly could be a snake den under there, but I wasn't too keen about searching for it. I walked gingerly around in the yard, thinking about the problem. It might have been just the one little moccasin that had bitten Anais — the one that I had draped across the chain link fence and was now shriveling in the sun — or there could be more.

Each night for several nights I went on snake hunts in the yard with a flashlight and hoe. I walked carefully, slowly. The first night I found another small, gray-green moccasin and I racked him smartly with the hoe. I killed a third one the next night. Then no more.

Anais had heard, somewhere, of a pro-life organization called Save-a-Snake. They were conservation-minded folks who would come to your house or farm and catch — and save — any snakes you didn't want around. We agreed they were just the ones to go crawling around under her house. We gave them a call, and the young woman on the phone said they would try to be out within the week and thank-you-folks-a-lot.

Anais slowly returned to good health and, as the workaday world kept on going beyond the rim of the river trees, we did nothing that did not please us. We joked about being so degenerately unproductive, so totally and unrepentantly white-trash *sorry*. At noon, after pouring glasses of white wine, we would fix cheese-and-jalapeño nachos and put them in the microwave; then we crunched on them as we sat at the big cluttered kitchen table and listened to Ella Fitzgerald tapes — watching Sarah, the small gray cat that sat on the microwave and watched us back.

If Anais would go into the bedroom afterward to fold laundry, I would come to the doorway, lean against the frame, and say that the bed certainly did look nice in its two o'clock shadow. And with the sun blazing down and the air-conditioner whirring — and the dogs asleep on the front room rug and the chorus of elm-tree locusts rising and falling outside like evangelicals of summer heat — we would take off our clothes and once again be as luxuriously, as deliciously *sorry* as we had been all the days before.

It was on one such afternoon that the Save-a-Snake couple drove out. Anais and I were in bed. She was in her climax, yelling out her curious cry, "Oh, *God*-dy, oh *God*-dy!" when the dusty, beat-up brown van stopped in front. I had finished — was arched back, almost ready to collapse and fall across Anais — when I saw them through the window. "Be with you in a minute!" I yelled as the door of the van opened and a tall, rawbony young man in overalls and a painter's cap got out. I held Anais, telling her in a rush, "The snake people are here!" and rolled off to my feet. I put on my shorts while Anais lay there grinning up at me (— grinning the way I liked to see her best, without lipstick, stretching the sensitive skin of her full and beautifully symmetrical lips). She eased off the bed, put on a shift and sandals, and we went out to meet the snake couple at the gate.

The rawbony hunter crawled around underneath the house to his heart's content but did not find any snakes. Anais and his plump companion talked about canning tomatoes.

One early morning in July we took our coffee cups and sat on a bench in the back yard beneath the trees, watching our bantams — the six we had bought at a flea market near the interstate. The garden we had started was now burned to a crisp, but in the little chicken run we had our six bantams to look at and enjoy and it was like money in the bank. The dogs roamed about in the yard, crapping huge craps; the river sycamores loomed against the sky. Anais and I sat close on the bench, drinking our coffee, watching the chickens go after bugs as we would have indulgently watched children at play.

At night we sometimes played Scrabble on the front room rug. Typically, I would find myself staring at one of the strange-looking words she had just made. She would casually slide each square into place on the board; then not smiling, but almost, she would lower her head and look at me over the top of her glasses: "Care-to-look-it-up?" she would ask. I would reach over and get her huge brown Oxford dictionary from behind the chair. I knew it was useless to check, but surely — surely *once* she would be bluffing. But there it was: a three-letter word meaning "Japanese urn." Or maybe she had made a two-letter word: "si." It would of course turn out to be the fifth note on some obscure scale or other. Through her curling cigarette smoke she would watch me close the dictionary, shake my head, and write down her points.

We played half a dozen games during the summer — each time at her suggestion, of course — and she won them all.

Despite Scrabble, I felt very much at home in the little white house. I did nothing in particular — wrote a little, read a little — and doing nothing was quite enough. Just being there with Anais was satisfying.

Yet there were times when I felt cut off from much of who and what I had been before I met her. I felt torn between the old framework of my life and this new, separate, almost hermetic world along the river: the flat, coastal southside with its fields, sunflowers, mesquites, dust-layered weeds; the canopied twilight mood along River Road; the river-bottom greenery, the almost undersea shadow-and-depth of trees, trees, more trees. . . . I had wandered into a strange and compelling oasis — with the sky, the familiar world outside seeming vaguely unreal — and I could not quite decide how to feel about it.

Sunk into the privateness of myself, I would rake leaves in the side yard — the mindless raking letting my thoughts unwind. It was obvious that Anais wanted me but I did not — so soon after my divorce — want to be wanted quite that much. I liked to be with her, but on my own terms. It was as if we were two boats that had drifted together and were rocking side by side in a pleasant, inland bay — with a rope flung loosely from one deck to the other to keep us from ebbing apart. I was definitely glad to have dropped anchor for a while, but I wanted to keep my freedom: to be able to pull up my anchor if necessary.

So I kept resisting, for the first month and a half, being pulled out of my own separate space. Anais was magnetic, but there were things about her that gave me the willies. I just couldn't cross the last hurdle of commitment until I knew, for sure, that I loved her.

Late one afternoon I asked her if she wanted to leave the house for a while — go have a drink some place, just get out. No, she said, she was feeling tired – running a 99.9 temperature again. I said, I think I'll go on the southside for a while, be back after a bit.

I left, bought a few home-grown tomatoes and a newspaper at a little roadside store, read the paper in the car under a tree. I drove in past the interstate and had a beer in a tavern that had a long, uncrowded bar and played *conjunto* music on the juke box. Afterwards I sat in my car in the dusk: just sat, like a rock enjoying its own inertia. I looked at cars passing, at the night falling. I decided that Anais would not want to fool with supper – she would just want to sleep as she had the afternoon before – so I got a bite to eat at a nearby cafe.

About ten o'clock I went back to River Road. Anais was wild. She was walking her slow walk in the front room, and smoking. She wanted to know who She was – this woman I had gone in to see. I told her – tried to tell her – that I hadn't seen anyone, that I just needed to get away for a little solitude and self. I opened the sack from the car and handed her the three glossy tomatoes. She threw them back at me and turned away.

It took two more hours of constant talk before finally – or so she said – she believed me.

II.

Nagging problems had been there all along but they were in the background – small stuff, I thought.

For one thing I could not stop feeling like a guest. It was Anais' house, her dogs, her lifestyle, her repair work that needed to be done, her friends that we went to see occasionally. It was her this and her that, and I understood it should be that way. I could not presume that I should walk into her life and take over what was hers. I did not want to act like the male chauvinist that she so stridently disliked. Things might have turned out differently – might, I say – if she had come to live at my house, in my town, among my friends; or if we had gone off to some neutral area to our house and started our life together. But the River Road dynamics were loaded: she was in her element and I was not. Thus at times I knew I came across to her as a guy who didn't seem to have much to offer or to say.

The truth is, I simply like to listen while Anais talked. I listened because she was interesting to listen to. I was also more passive than usual that summer because at age 52 I felt that I did not have a whole lot more I wanted to say to people. I still said on paper what I wanted to say, and for nine months of the year in school classrooms I talked more than enough. And I enjoyed talking to my friends back home. But there on River Road, in the presence of someone like Anais, I was content to listen in the same way I would be content to *look* at Rio de Janiero: simply because Anais was fascinating, perhaps the most unique person I had ever met.

So I listened and learned. I was a spectator at her show. (Not once, however, did I get the impression she was interested in learning about me, or about anyone else. She was basically self-absorbed.) In a way Anais was always on stage — for an audience of one or for a roomful. She liked being witty and having eyes turned toward her. She was delighted to be delightful — and unconventional — and she knew how good she was at impressing people. That was what was so disconcerting about the "Damn I'm Good" button she would sometimes pin on her longsleeved work shirt and wear around the house. As a Renaissance Woman who could do almost anything she set her mind to — from learning Latin to changing the oil in her truck — she was the last one who needed to announce — to proclaim — her worth. Yet, as I began to see, beneath her bold air of female liberation and self-assurance was a person who did not seem to believe in herself at all.

(She was mechanically inclined, like her father. In contrast, I am not handy with my hands, have no ability to construct things. She could look at a schematic drawing — "Place Gadget C through Flange 12. . . ." — and it was clear as spring water to her. In early June we — that is, she — put together a portable typing table she had ordered for my birthday. We sat on the living room rug among the screws and bolts and from time to time she would glance at me through her curling cigarette smoke — smiling at my discomfort. No Leonardo, I.)

As the weeks went by on River Road I learned a little about her childhood, somewhat more about her teenage years, significantly more about her marriages and male involvements. Her family, of course, cast the shape of her life. Anais' mother — a gifted mathemetician who endured, with great bitterness, years of teaching high school algebra in San Antonio instead of doing research in higher mathematics at Berkeley — felt she had married beneath her. Anais' father, a solid citizen from a German farm family, apparently agreed with his wife's assessment. He preferred working long hours as the head mechanic at a car agency to facing silent accusations at home. But it was Donald, Anais' mildly retarded older brother, who turned family drama toward tragedy. In his mother's eyes Donald was not lesser but more; flawed and thus more deserving; the one male in the house who could do no wrong. Anais? She threatened her sullen, aggressive brother by her very existence. Fiercely, the mother gave Donald her attention, protection, encouragement, love. Anais, like a Cinderalla, was denied them all: Mother punished daughter for the sin of being so conspicuously bright. (Anais would go through life cowering before the private image of a disapproving mother; the future Mensa member would never be good enough for Mama . . . .)

At 16 Anais, who was brainy and pretty and trying desperately to look and act and be sophisticated, met a fellow who was stationed in San Antonio with the air force. He was in his late thirties, bored, heir to a family fortune back in Pennsylvania. He took Anais — smoking her first cigarettes, wearing her first high heels, winning her first prize in an art show for a brooding self-portrait — and did his own Professor Higgins on her: giving her a bored gentleman's culture course while also getting her pregnant. Later that year she had her first abortion. (Needless to say, it did not please Mother, and Daddy never found out.)

At 17 she had graduated from high school, entered college, and tried out the first of her four marriages. ("I had almost forgotten about him," Anais said as she talked about her college days. After a freshman history class a tall fellow from the back row came up to her and asked if she wanted to get married. "I told him, 'why not?'" and they went across the border to Mexico for a quick ceremony in Piedras Negras. They stayed together for six months before divorcing. She taught him fencing.)

She was 20, working for an advertising agency, when she married Ted, who was older and – by her assessment – brighter than she was: perhaps, she thought, the only man she ever met who was. That seemed to be his hold on her: he could treat her any way he wanted to and she would take it because he was brilliant. And handsome. And also homosexual, or bisexual, except she did not realize it for a while. They remained married for 25 years even though he was gone much of the time – he did some kind of secret work for the government – and even though he treated her with a kind of grand indifference. They lived in Miami and Nebraska and for a while in the Philippines. Much of the time she was alone so she painted watercolors and wrote haikus.

While they were living in California she did advertising work for a business concern – apartment complexes, shopping centers – that she later found out was controlled by the Mafia. (I have, in a box, her 200-page manuscript about that experience.) She also entered law school and took classes for a year.

Her classes and her marriage to Ted came to an end one weekend, she said, when he tried to kill her. He hit her from behind in their living room and then left, assuming she was dead. "Unfortunately for Ted," she told me as she worked on a crossword puzzle, "I have a German's hard head." She suffered blackouts and spells of amnesia and ended up in the hospital. Ted had moved to an apartment in another part of San Francisco. When she got out of the hospital she recuperated at home after first stealing back from Ted's garage their red Porsche she now considered hers. She phoned Ted, telling him what she had done and daring him to do anything about it. I could never understand why she had not pressed charges for assault or attempted murder. She was never clear on that point. I think she granted Ted-the-genius the right to do whatever he wanted to do – up to a point.

A few weeks later a girl in her 20's was found dead of a drug overdose along the shore below Anais' house. The girl's father, Clarence, lived alone across the street from Anais, and they were soon offering each other consolations for their various losses. Kindly Clarence seemed a good-enough sort to Anais – he was quite good at crossword puzzles and knew a lot about the stock market and investments. She was without money or resources – and still had blackouts, severe headaches, memory loss – so they decided to get married. They sold the Porsche, bought a second-hand trailer house, and hit the road to Texas. She lived with him in the trailer house for a year – on family farm property about 30 miles south of San Antonio – before learning the bad news: both Clarence and his brother Eugene, who lived in Atlanta, were con artists who operated throughout the South and Southwest and were wanted by federal authorities. (They were part of Chapter 10 in a book about swindlers that someone sent to Anais. I read about their remarkable exploits as I sat out under the elms.)

End of Marriage #3: Anais divorced kindly Clarence and moved to River Road.

(I do not know what the male sexual organ meant to Anais, but apparently she needed the presence of it, as she got older, in a way that was excessive; yet at the same time she was constantly on guard against it: *it* meaning male domination.

Everything about her was contradictory. She seemed to seek out brutalizing encounters – hitchhiking in California after her divorce from Ted she was picked up by two men who almost dutifully raped her. But she was constantly on the lookout for sexism: she saw it lurking everywhere and was ready to defend herself against it. And I can understand why. I can visualize her years in the advertising world when she had to defend herself against the crudest of male bosses and clients: defending herself with the formidable array of her tongue and intelligence; defending herself as she had been doing since childhood. A lesser person, a lesser defender, would have soon gone under. But she had obviously paid a price. She looked for offenses against her where none existed. She was suspicious, volatile, accusatory, wrongheaded. I can't recall a single time when she admitted that she had misjudged or erred.

I read once about Bubu de Montparnasse, a prostitute in Paris who submitted to men who beat her because, she said, they were like a "strong government" who could also protect her. . . . I suppose Anais, more than anything else – despite her bold, intimidating posture before the world – wanted to be protected: wanted security as she began to look down the barrel of her own mortality.

I cannot believe Anais was true-blue to Ted for 25 years, but she said she was. They seldom had sex and then not lovingly; they often stayed in separate rooms of their house. . . . After her marriage to Clarence ended she said she would go into a highway dance place near the farm, sit in a booth while she drank a beer and looked over the available males. When she had finally picked out a suitable-looking candidate, she would signal the waitress who would then take over a drink to the prospect with Anais' compliments. He would come sit in the booth with her for a while, then off they went to the trailer house for the next several hours or for the night.

I learned that until I came to River Road she had lovers from all over the county who made her little white house in the elms their pit stop at least once a week. Usually, she told me, she couldn't wait until they were gone: after the sex was over they bored her.

. . . Need them, then cast them out. It was as though sex to her was a way for her emptiness to be filled up, a way to be reassured she was still wanted, still desirable.)

In early August we went back to the flea market and got six more bantam chickens. Then Anais decided she wanted milk goats, too, so we went back and got three Nubians. We already had cats, and fish, and a small rabbit I found one night in the yard. And greyhounds.

When Anais started planning the construction of a backyard room for the prospective greyhound pups, she got the name of a guy who did part-time construction work on the southside — a Ray somebody. He came by one afternoon — smallish, dark-haired, a smoker — and gave her an estimate. She walked around the proposed backyard plot — coughing, feverish again — and discussed cement mixers and window spaces while I tried to get some materials ready for new classes I would soon be teaching.

And suddenly we were out of sync. She was increasingly hostile; weekly confrontations were becoming more and more destructive. I thought: it's going to be this way forever; there will be no change. Rather than live like strangers under the same roof, arguing, I left.

Twice I packed my stuff in my car and left River Road. And twice, at a hot 7-11 telephone booth on a hot San Antonio street, I called

Anais and we talked and afterwards I drove back beneath the arch of trees and unpacked and hoped that it was the end of the quarreling. Each time I was glad to be back, each time the prospects looked okay, and each time the peace only lasted a week.

By then I knew I loved her. I also knew that what was waiting for me outside her half-acre was the chaos of my mid-fifties with all my personal bridges washed away. I had found a little island of satisfaction by the river, a place and a person that suited me, and I did not want to leave.

I kept thinking: Maybe if we just take our time, going slow over the recognizable rapids we now know are here, we can make it into October and watch the leaves fall; and then into winter and watch the outside world from within the snugness of the house, sitting by the bedroom stove and sipping rum-and-Coke. We'll make us a contract – a commitment – to stay together at least until spring and then we'll see. . . .

But in late August, for the third time in three weeks, I left and did not return. I got an apartment on the north side and started living – if it could be called that – away from River Road.

(I had walked across the bedroom to where she was watching me pack. She stood by the wall in her tight blue jeans and boots, smoking. I put my arms around her and held her. Her body was stiff, unyielding; her skin was cold. When she finally spoke it was like a statue talking: "All I'm good for, you know, is a lay."

She only half meant it, but the half she did mean was a terrible summing up.)

I called her in September on the first day of school. She talked a bit about the construction of the room, about Apollo and Cleopatra. Then she told me – almost casually mentioning it in passing – that she was probably going to get married the following weekend.

Married?

Yes, she drawled. To Ray.

(Random snapshots of River Road:

–The second day after I had moved in she walked behind me in the kitchen, ran her finger down the back of my shirt and said, "Youwanna-get-married?" I remember thinking: My God, I don't even know you. . . . I *want* to know you, but *marriage*? After just a *day*? You're crazy.

—"I've never wanted to fix breakfast for anyone before," she said one morning as we washed the breakfast dishes.

—She would drift by the kitchen table where I was reading or writing and leave me a note on a piece of paper: "Thou swell."

—One night, after a reconciliation, we stopped at a Whataburger. I went inside, stood by the counter and waited. I would look out to the cab of the pickup where Anais — wearing her gangster hat — waited with her friend's monster-sized dog, Lewis, that we were keeping for the weekend. Lewis filled up almost the entire front seat and drooled. We went down a back street, parked, ate our hamburgers on a grassy rise of ground. The night wind was blowing in from the coast, and tree shadows flowed back and forth across the sides of houses behind tall street lights. Anais smiled at me as she sucked her milk shake through a straw and Lewis moaned in the cab.)

III.

I tried to write lesson plans for school, to read, to put away in the apartment the few odds and ends I had brought with me from River Road, but during the long hours of the weekend I couldn't concentrate on any one task for very long. So I sat down at the typewriter and started a letter to a friend of mine who had visited River Road during the summer. Ostensibly it was a letter, but it was actually just a chance for me to analyze and re-analyze on paper what had gone wrong: excusing Anais, then blaming her:

". . . One afternoon in late July, when we were making love, I told her, 'Don't look now, but, by God, I *love* you.' It came out easily, spontaneously; the words were just there in my mouth. And I was so damn glad it was finally over — the long waiting for some kind of genuine, gut-level acceptance of this strange and amazing woman.

"It's weird, but if I had to point to a single moment during the summer when our relationship started to go sour — when Anais began to pull away from me — I think it would be that one: when I finally told her that I loved her. It was as if I was no longer someone who had 'power' over her. She had got her own power back — her own self-possession — and I was like one of the guys at the dark restaurant place I mentioned: someone to experience and throw away.

"You see, I think Anais has been hurt so many times before — plus five miscarriages — and tried so hard to find some kind of happiness that by the time I met her she was no longer interested in Romance. She wants *marriage* — the badge, the official seal and permanence of it. It would be fine to have it again with someone she could respect, but if not, well, so be it: she'll take it any way she can get it.

"I keep wondering at what cost she maintains her air of unflappable self-assurance: the cost of her life? She would rather break than bend. She will never say, 'Help me, I'm messed up like everybody else; my whole life is a testimony to being messed up.' Because she refuses to be like anyone else. And she refuses to quit smoking. She coughs and lights another cigarette and coughs and smiles sort of coquettishly and coughs some more.

"And how about this: All during the summer we would each buy groceries and stuff when we were in town, but we had agreed that at the end of the month we would settle up the other expenses — phone, electricity — and split the costs. And I did that for June and July. But when I was leaving, finally, a couple of weeks ago Anais sat at the table, smoking up her Marlboro storm, and she totalled up my 'August and final bill.' She charged me an additional $600 rent for the three months — my 'half' of a 'typical $400-a-month rent.' That's Anais for you: charging her ex-boyfriend and ex-lover summer rent.

"She's supposed to get married this weekend to a guy who is building a backyard room. She's known him a total of two weeks. (The first afternoon that he came out, slouching around in the back yard with his measuring tape, Anais told me as she watched from the window: 'I could have him if I wanted him.' Sexually, she meant. I didn't even bother to answer. It was just more of her bad-ass bluster.) What I think is this: she doesn't want to go back, now, to the side-of-the-road restaurant stuff — the casual whoredom of past times. So she's going to turn herself over to this Ray joker and have him keep her life from destroying her. Something like that.

"I'm here in San Antonio at the edge of Alamo Heights and the edge of my self-control. It's Sunday afternoon and that's not a good day any more."

I intended to mail it but got sidetracked and put it in a folder with some pictures of Anais. I remember thinking, finally, that it really wasn't a letter but just an exercise in writing-it-out therapy for me – and certainly not anything I should send on to my friend Carl, who was having his own problems that year. I decided to write him a less personal, more upbeat report when I could summon up the energy.

I moped my way through the fall, avoiding southside streets and trying not to think of Anais and Ray. . . . I didn't see how anyone could last much longer with her than I had – before she drove him away for reasons too complicated for mere mortals to unravel. I figured Ray for an opportunist out for any bird's nest on the ground.

I gave them until New Year's.

My estimate was off, but not by much. I got a call from her in early March. It was Sunday about two o'clock and I was lying around the apartment trying to get rid of a cold. The telephone rang, and Anais offered her "What's-going-on" as nonchalantly as if we had just had a buddy-buddy chat the week before. "Not much," I said, "and you?" "Getting by," she said.

After a few more preliminaries about dogs and ailments – she had been sick with a bad case of pneumonia in January – I finally asked about Ray. He had gone back to Michigan in February, she said, and good riddance.

We maintained our low-key oh-yeah?-is-that-right? casualness until it gradually gave way to a Well-if-you-want-to-I-want-to honesty. Meaning, now that six months have passed, what have we learned and what do we want?

I told her I had a cold but I would meet her for coffee some place near the interstate.

When I got to Cap'n Jim's Cafe Anais was waiting in her truck with Cleopatra. She looked pale and thin. We talked for an hour or so in the cafe, and I listened hard – alert to every nuance of her manner and voice, almost trying to read the very air around her. She was agreeable, forthright, complaisant – as if properly tempered by her last marital adventure although not humbled or apologetic. It was as if the Good Anais had been in battle with the Bad Anais and had won.

We talked about the idea of my coming back to River Road, and we both seemed to say: All right, but no need to rush. We can take it slow. We've been through trying times.

And Ray? I finally asked.

A reprobate, she drawled. She said she had already filed papers for divorce. I didn't press further. (Later, after I had moved back to River Road, Anais went on binges of Ray-bashing, ridiculing everything about him from his belief in reincarnation to his taste in socks. He had gone off for weeks at a time without letting her know where he was. She had advanced him money when they married and he had never paid her back. She had also bought him a new jeep because his Pontiac was junk and he needed, he said, something to drive while he looked for work. He never found a job. "Sounds like he really *was* 'sorry'," I said, catching her eye, and she giggled — actually giggled.)

Each afternoon during the following week I drove south to River Road after school. Anais and I would walk down to the river, hand in hand; then I would drive back to my apartment for the night. I was determined to take things slow, do the job right. I kept on the lookout for any sign that might show a return to "August tendencies" — the suspicion and paranoia of those intolerable weeks. I found nothing but pleasure and delight.

In mid-March I moved back with Anais, and we had one good week of almost honeymoon-like tranquility. We took baths in the old tub together. She gave me a haircut out under the trees. I bought her a baby orchid corsage — "just on general, middle-class principles," I said — and she told me it was the first flowers she had been given since high school.

She called me several times at school to update me on the latest escapades of the greyhound pups. At night I would be grading papers at the kitchen table and I would feel the urge to call out to Anais: "I love you." Maybe an hour later, or the following night, she would appear in the doorway, light a cigarillo, and say it too, nice and spontaneous: "I-love-you."

And then her sickness started. First it was fever and a hurting in her chest. We went to a southside doctor who gave her antibiotics. She stayed in bed almost all day every day. She kept running a temperature and began feeling worse.

I checked her into Humana Hospital, where doctors took x-rays and ran tests. A specialist thought she might have a rare lung fungus, Desert Fever, which is difficult to treat. After a week Anais was released with watch-carefully-and-wait precautions.

That was in early April. She got up each morning for a while, went back to bed and stayed for the rest of the day.

In mid-April I got a whiff of it: "the August tendency." I heard it one morning in her voice at breakfast before I left for school.

(Anais had three voices. She used one voice when she was attentive, modest, cooperative. It was light, somewhat breathy, almost excitable and girlish. Her second voice was the drawling, hypnotic one: low and slow. It was her resonant telephone voice to friends, lovers, strangers she wanted to impress. It contained a trace of the Germanic brogue of her childhood. Her third voice was lower than the other two – throaty, hard-edged, and with a longer drawl. It was her In Command voice, and chilling to hear. She would have used it had she been a queen or a despot.

All three voices reflected the inner Anaises.)

At breakfast I refused to believe that anything was wrong. Anais was sick – and sick of feeling sick. That was all. I fed the dogs and drove off to school. In the afternoon, I remember, I spent an hour or so going to three different record shops on the southside trying to find a new Pavarotti tape she would like. I bought juices and dog food and hurried home.

She was sitting in the front room when I got there. She was smoking and staring out past the butane tank toward the river. I put the grocery sacks down, went over to show her the Pavarotti tape I had bought.

"I read the letter," she said and kept looking out the window.

"What letter?" I stood there with the tape in my hand.

"Your newsy letter to your friend Carl."

Then I saw the folder on the rug and understood. She had been prowling through a box of mine and came across the unmailed letter

I had written in September. I didn't remember all that was in it but I knew it couldn't have made very pleasant reading for Anais.

I could have told her, "You had no right to go through my personal stuff" or "That was when we were both upset" or "My God, it hasn't got anything to do with us *now*." I could have said a lot of things. But I kept standing there with the tape, trying to recall the words I had poured out that Sunday afternoon, and looking at Anais, who was drawn and tired. Just at that moment a car stopped out front: friends of Anais who had heard she was sick and came to cheer her up with fruit and flowers.

Somehow I got through the amenities of greeting the friends and listening to Anais charming them from her chair – between coughing attacks – with funny stories. They stayed a couple of hours and I fidgeted around the house and in the yard.

After they had gone Anais went to the bedroom. I cleaned up the kitchen and went to clean up the uglier mess of the letter.

Except that the bedroom door was closed. I tried to open it, and it was locked.

"What. . . ." – is as far as I got.

"If you try to open that door," came her voice from the other side, "I'll shoot."

I had seen the gun. She kept it in a drawer by the bed. The previous summer she had taken it out in the yard and showed me how proficient she had become in using it. "Any son-of-a-bitch who comes around when he shouldn't will soon be a dead son-of-a-bitch," she said, shooting four or five rounds at a tomato soup can on a post.

"Anais," I said, "open the door and quit being ridiculous."

"You try to force your way in here, and this "whore" will see to it that it is both your entrance and your exit."

I frankly didn't know if she would shoot or not. I spent the night on the living room rug.

When I drove off to school the next morning the bedroom door was still locked. I called at noon, but she hung up on me. When I came back that afternoon Ray was hanging out clothes on the back clothesline. The jeep was parked out front.

To her, her position was clear: I had insulted her and she was righteously incensed. But I knew better: it was just a convenient excuse.

My reading of the "August tendencies" at the breakfast table had been correct and she had grown tired of me, needed a change in male scenery, whatever. She had obviously been telephoning Ray — for how many days before? — and was waiting for the opportunity to invite him back.

. . . It was a grim late-afternoon: Anais sat in the front room chair, smoking, her gun in her lap; Ray shuffled about. During the hour it took me to load my belongings into my car I did not speak to either of them, and they did not speak to me. They murmured now and then to each other.

With a final paper sack of odds and ends in my hand I stopped in front of Anais on my way out the front door. I looked at this woman that I loved and said the only words I could think of that seemed appropriate. "You're not 'sorry,'" I said, "you're scum."

She did not look at me or make a reply. I drove off, and it was the last time I ever saw her.

IV.

What I know of Anais' final days is based on random reports from friends of mine and acquaintances of hers. In June the Humana doctors removed part of her left lung. She had cancer, not Desert Fever. Apparently she then became addicted to the morphine which she took during and after the operation and spent most of August in a drug rehabilitation clinic in the cedar hills north of San Antonio. Ray was either dismissed from further service or left of his own accord — interpretations varied. He might have been understandably miffed when Anais took up with Burton, a fellow patient at the rehabilitation clinic. Burton, an alcoholic, was a huge man who wore a speech-assistance device and ran a salvage yard in Austin. Anais found him sexy and amusing and during September he was the next caretaker for her dogs on River Road. But he stayed drunk and proved to be a man with a violent and uncontrollable temper. Anais, finally scared, called the sheriff, who had a peace bond placed on him.

The ever-lurking Ray showed up again and Anais hired him to take care of her and the place — no bedroom privileges. He was with her when she died the following April. Her body was cremated.

V.

It hadn't mattered, for a while, if Anais was *lovable* or not. She was like a gyroscope, a fixed center, and she had drawn me to her. Somehow she had seemed unassailable: no matter which way she was thrown by adversity, she kept landing on her feet.

"Death," too, should have been just another of her bizarre but temporary setbacks. She should have reappeared out of a back room, paler than usual but no worse off: smiling with that knowing look of hers, head lowered, peering over her glasses.

. . . Several years after she died I came through San Antonio and went by River Road. I drove through the arched oaks, got out of my car, and stared in the mid-summer heat at what remained of her hideaway acre. It was like revisiting a battlefield after the war was over. Her house had burned down and the yards had turned to trash and weeds. A few straggly sunflowers grew where the front room used to be, and broken limbs lay in the square of the bedroom area. A tin strip bordered the kitchen foundation.

The locusts whirred wildly from the line of river trees, and they were like a chorus of destruction: Now both Anais and her little white house were ashes.